AIR SURVEY OF SAND DEPOSITS BY SPECTRAL LUMINANCE

OPREDELENIE TIPOVOGO SOSTAVA PESCHANYKH OTLOZHENII
S VOZDUKHA PO IKH SPEKTRAL'NOI YARKOSTI

ОПРЕДЕЛЕНИЕ ТИПОВОГО СОСТАВА ПЕСЧАНЫХ ОТЛОЖЕНИЙ
С ВОЗДУХА ПО ИХ СПЕКТРАЛЬНОЙ ЯРКОСТИ

AIR SURVEY
OF SAND DEPOSITS
BY SPECTRAL LUMINANCE

by

MARIYA A. ROMANOVA

Authorized translation from the Russian

CONSULTANTS BUREAU
NEW YORK
1964

ISBN-13: 978-1-4684-7205-9   e-ISBN-13: 978-1-4684-7203-5
DOI: 10.1007/978-1-4684-7203-5

Library of Congress Catalog Card Number 63-21214

The Russian text was published for the Laboratory of Aeromethods
of the Academy of Sciences of the USSR by Gostoptekhizdat (State
Scientific and Technical Press for Literature on Petroleum and
Solid Fuels) in Leningrad in 1962. The English version contains
revisions and supplementary material provided by the author.

Мария Андреевна Романова

Определение типового состава песчаных отложений с воздуха
по их спектральной яркости

CONTENTS

INTRODUCTION

This book gives an account of a method of determining the type composition of sandy deposits from the air by the nature of their reflection of radiant energy. The measured spectral luminance of a rock outcrop relative to the luminance of a standard is the basic property considered in this method. The spectral luminance factors ρ_λ are measured by means of special instruments – a universal photometer, cinespectrograph, spectrovisors, and other spectrometers.

The measured values of ρ_λ of the investigated rocks are interpreted geologically with the aid of some methods of mathematical statistics so that a means of lithological mapping of sandy deposits from the air can be found.

The importance of developing a method of determining the type composition of rocks, primarily sandy deposits, from the air stems from the extensive occurrence of deserts on the earth's surface. Sandy deserts occupy vast areas in the southern latitudes of the USSR and occur in western and central China, Pakistan, Iran, India, and other states. In these deserts there are numerous deposits of petroleum and gas, and a search for water is continuously in progress. Successful prospecting for the above minerals depends on a determination of the material composition of the sandy deposits in these deserts and the solution of several questions of recent tectonics, paleogeography, and quaternary geology, the answers to which might also be given by a study of the distribution of lithological types of sands.

Ground methods of investigating the lithological composition of sands entail the conduction of work in extremely arduous conditions, where travel in any direction requires a great expenditure of effort, time and money. The replacement of ground methods of geological mapping of sandy deposits by aerial investigations (aeropetrographic mapping) would very considerably reduce the cost of the work and the time required for its conduction.

An investigation of the reflecting properties of sandy deposits showed that they belong to the simplest type of objects as regards the way in which they reflect radiant energy – they give diffuse or orthotropic reflection. This simplified the photometry technique to some extent, and we did not need to use such devices as polaroids in our spectrophotometers nor to ensure very accurate lines of flight of the airplane in the aerial photometry of outcrops.

Thus, the results of investigations of the spectral luminance of sandy deposits for the purpose of geological mapping of present deserts can provide a starting point for the development of a new method of geological research – aeropetrographic mapping of rocks. The further development of the method will entail the improvement of spectrophotometers and the design of new photometric instruments equipped with polaroids and with computers. Progress in this direction will make it possible to determine from the air not only the type composition of sandy deposits, but also the composition of any rock. The importance of the solution of this problem for geology and cosmogeology is obvious, particularly in the light of the recent achievements of Soviet science.

In accordance with the proposed aim of this work, the material presented is arranged in the following way.

The first chapter, which serves as an introduction, explains the relationship between such characteristics as the color of rock and its spectral luminance. This decided the line of research to be followed in the development of the method.

The second chapter is devoted to the basic concepts of photometry, since many of these concepts are not commonly encountered in the work of a geologist. The exposition of the photometric concepts is based on the generally adopted system, so that the meaning of the measured spectral luminance factors is clearly brought out. The material in this section is taken mainly from the work of Landsberg (1957) and the results of some investigations of the diffuse reflection of finely dispersed media (Girin and Stepanov, 1954). In addition to the definitions of the main photometric properties of objects, this chapter contains illustrations of the types of indicatrices of the luminance factors of some rocks of practical interest.

The third chapter gives descriptions of the apparatus and instruments which we have used for the photometry of geological objects. At the beginning of our work there was only one photometer (the FM-2 universal photometer) really suitable for measurement of the spectral luminance factors of rock samples. The development of observations of the reflecting properties of natural objects and, in particular, the need to carry out photometry of these objects from the air necessitated the construction of special aerial spectrophotometers.

The Laboratory of Aeromethods has now developed a whole series of spectrographs and spectrometers, some of which have undergone repeated testing in the spectrophotometry of sandy deposits, forest vegetation, or water surfaces from an airplane or helicopter. In the compilation of the third chapter the author enjoyed the generous collaboration of K.E. Meleshko, to whom she expresses her sincere thanks.

The fourth chapter deals with the methods of treating the observations of the spectral luminance of geological objects and the methods of converting from reflection spectra recorded by various methods to curves of spectral luminance factors.

The fifth chapter gives an account of the methods of numerical evaluation of these curves and the methods of comparing the results of photometry of rocks with their particular petrographic and mineralogical composition.

The methods of checking the relationship between the photometric parameters of an investigated rock and its content of characteristic components lie in the field of mathematical statistics and, hence, some statistical methods of calculation, illustrated by specific examples, are given at the end of the fifth chapter.

The sixth chapter is devoted to the results of laboratory observations to determine the relationships between the reflecting properties of rocks and their facies characteristics. These results were obtained in the investigation of a series of samples of sandy-silt* composition collected by the author from Middle Jurassic, terrigenous Cretaceous and Tertiary deposits. In each of these cases the samples were collected from sections which had been formed in conditions typical of the platform type of sedimentation or in conditions close to geosynclinal. The main aim of these investigations was to discover the relationships between the spectral luminance of rock and the conditions of its formation and to find a means of determining the facies characteristics of sandy-silt rocks from their spectral luminance.

The seventh chapter gives the results of aerial mapping of the lithological composition of sands and the results of the first experimental aeropetrographic survey. Practical aeropetrographic surveys by photometry of sand outcrops from the air could not be undertaken until the laboratory work on the photometry of rock samples had been successfully completed and the existence of a relationship between the type petrographic composition of the rock and its spectral luminance had been established.

The laboratory stage of development of the method occupied the period from 1953 through 1956, and aeropetrographic mapping was begun in 1957. Systematic experimental aeropetrographic surveys were carried out in 1957, 1958, and 1960.

Throughout this work the author was generously assisted by A.B. Vistelius, N.G. Kell', and K.S. Lyalikov, to whom she expresses her sincere thanks.

---

*In Russian terminology, "silt" is the term used for size 0.1-0.01 mm; "sand" is used for coarser materials (0.1-1 mm), and "mud" or "peltic material" is used for finer materials (< 0.01 mm) – Publisher's note.

CHAPTER I

COLOR AND SPECTRAL LUMINESCENCE OF ROCK

Since the method being developed for the determination of the type composition of rocks from the air is a new geological method, we will have to consider some aspects of the history of this question and the different physical definitions involved in the investigation of the reflecting properties of rocks.

For a reliable and objective evaluation of the reflecting properties of rocks under investigation, the characteristic selected for measurement must be stable and very easily determined from the air. For the selection of such a characteristic, a relationship between the color and spectral luminance of rocks was determined. This decided the rational line of research to be followed in the development of the method described below.

The color characteristics of rocks and different minerals have long attracted the interest of many investigators. From ancient times the human eye has been accustomed to distinguishing rocks from one another primarily by their color. At the start of our investigations on the reflecting properties of natural objects, the color of the rock was adopted as the basis for the measurements.

The average human eye distinguishes colors from one another by their hue, purity, and luminance. Of these characteristics, the hue and purity of the color constitute the chromaticity of the object. The hue is determined by the wavelength λ of monochromatic radiation which must be added to white light to give the particular hue. The color purity p is equal to the ratio of the luminance of the monochromatic component B_λ to the sum of the luminances of the monochromatic component B_λ and the luminance of the white component B_W. The sum of these luminances $(B_\lambda + B_W)$ is equal to the total luminance B_φ.

It has been experimentally established that any color can be obtained by mixing particular amounts of any three mutually independent colors, i.e., colors which cannot be obtained by a mixture of the two others. It follows from this this that three parameters are necessary and sufficient for the definition of any color. Besides this method, another method of color definition, based on the mixing of white light and any particular monochromatic radiation, is widely known in practice. But if white light is regarded as a mixture of two complementary monochromatic radiations, the second method of color determination reduces to the first, i.e., to the mixing of three mutually independent monochromatic radiations. In some cases the color parameters can be distributed so that two of them determine the chromaticity of the radiation and the third determines its luminance.

Of the three existing basic colorimetric systems ($B\lambda p$, RGB, and XYZ), the $B\lambda p$ and XYZ systems are the systems which have mainly been used in practical geology.

The $B\lambda p$ system is based on the obtention of any color by the mixing of white light with monochromatic radiation. Here B denotes the luminance of the color, λ its hue, and p the purity of the color. The $B\lambda p$ colorimetric scheme is simple and gives a clear representation of color but does not allow color calculations.

The RGB system, or the trichromatic system of colorimetry, is based on the above-mentioned law of color mixing. In this system the three mutually independent colors adopted as basic are red with wavelength 700.0 mμ (R), green with wavelength 546.1 mμ (G), and blue with wavelength 435.8 mμ (B). These colors were adopted as unit colors by the Commission Internationale de l'Eclairage (C.I.E.) in 1931. The RGB system can be used to solve many problems of colorimetry, particularly the determination of the color of reflecting surfaces if they are illuminated by radiation with a known spectral composition; but the drawback of this system is the complexity of the color calculations due to the occurrence of negative coordinates. A third colorimetric system — the XYZ system — has been introduced in order to eliminate this drawback.

The XYZ system. The primary colors of the XYZ system are not actually realizable colors. The whole field of real colors lies inside a color triangle and any color in the XYZ colorimetric system is expressed by the equation

$$C = x'X + y'Y + z'Z,$$

where x', y', and z' are coefficients of the color C and represent the number of unit colors X, Y, and Z. The XYZ system is constructed so that the luminance of the color is equal to the coefficient y'. The coefficients x and y in principle are independent variables, but, as experimental observations have shown (Vistelius et al., 1954), there is a strong linear relationship between them in the case of the evaluation of the chromaticity of rocks.

All color calculations are based on the law of additivity of the color coefficients when colors are mixed (the coefficients of the color are equal to the sums of the coefficients of the mixed colors). The law of additivity of the coefficients of a mixture is a confirmation of the experimentally established fact that the color of a mixture of several radiations depends only on the colors of the mixed radiations and does not depend on their spectral compositions. It would be possible to derive an infinite number of spectral luminance curves obtained by photometry of objects producing the same color sensation in the average human eye, and the shape of these curves would be quite different. An example of this is shown in Fig. 1.

In the majority of cases the evaluation of the chromaticity of rocks is of a subjective nature, and it is rather difficult to demonstrate its physical sense. This difficulty is accentuated by the properties specified by the laws of color mixing:

1) For any monochromatic color it is possible to select a second chromatic color which on optical mixing with the first in a particular quantitative ratio gives an achromatic color (white, gray, or black).

2) Optical mixing of noncomplementary colors gives colors with a hue intermediate between those of the mixed colors.

3) Colors which appear absolutely identical give the same results in optical mixtures, irrespective of the physical composition of the light fluxes causing the sensation of these colors.

Probably the first objective evaluations of the color characteristics of rocks were the investigations of the Soviet geologist Popov, who used Ostwald's color atlas for the measurements. A further development in the quantitative evaluation of the chromaticity of rocks can be found in a resolution of the U.S. Sedimentation Committee, which recommended the use of Munsell's color solid for color characteristics. However, the use of color atlases does not secure precision in the evaluation of the color characteristics of the investigated rocks, since in almost every case elements of subjectivism creep in.

One of the first investigators to employ instrumental measurement of rock color was the Moscow lithologist Danchev, who over a period of several years measured the color of sedimentary rocks and used this characteristic as one of the indices of the conditions of sedimentation (Danchev, 1946, 1947, 1956). Danchev's technique was roughly as follows:

a) the rock sample was ground to a powder, which was then compressed, and the obtained briquet was measured on a Pulfrich photometer;

b) a set of standard gelatin filters with wavelengths 460, 480, 510, 540, 565, 630, and 650 mμ was used in the photometry;

c) the color of the measured rock sample was quantitatively characterized in accordance with Ostwald's system, but the color components were interpreted in terms of the amount of particular constituents in the rock.

For clarification it must be remembered that Ostwald, working with a half-shadow photometer of his own design, used various filters. He proceeded from the view that the test surface on illumination by monochromatic rays of the corresponding color would appear white, since a filter of the corresponding color would only transmit rays which gave a maximum reflection at the

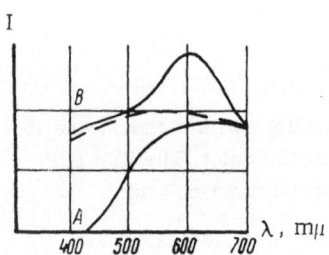

Fig. 1. Two radiations differing in spectral composition but producing the same sensation of yellow color. The broken line gives the spectral composition of white light (from Volosov and Tsivkin, 1960).

4

particular surface. The same surface illuminated by monochromatic rays of the complementary color would appear black, since a filter of the complementary color would transmit minimally reflected rays in the same conditions.

The difference between the maximum and minimum reflections, according to Ostwald, gives the fraction of the pure color of the test object. Thus, in his classification of colors Ostwald used the formula

$$W + S + V = 100 \ \%,$$

where W is the brightness, or white content, corresponding to the lowest point of the spectral luminance curve; S is the black content and corresponds to the length of the ordinate from the maximum of the curve to the point taken as unity; V is the color content and is equal to the difference between the maximum and minimum values of the spectral luminance. According to Danchev's interpretation, these values correspond to the amount of particular constituents in his investigations of sedimentary rocks.

Danchev's published data show that the quantities which he used lend themselves to some extent to geological interpretation and are of an objective nature, but his technique as a whole prompts the following comments:

1. The pulverization and subsequent briqueting of the rock sample make the determination of the hue more definite, but the reflecting properties of the rock and its hue may be considerably altered. Moreover, such a procedure is quite unsuitable if the extensive use of color for aeropetrographic surveys is envisaged.

2. The standard gelatin filters used in the work do not provide accurate measurements of ρ_λ, since these filters have wide transmission bands and low transparency.

3. The system of color measurement devised by Ostwald for specific purposes does not give a representation of rock color in Danchev's interpretation. According to the method adopted by this investigator, the color of rock is determined from the maximum difference between the values of the spectral luminance of the rock measured in certain fairly wide regions of wavelengths. In this case the spectral luminance curve with all the bends characteristic of the test object is not utilized at all.

As the above comments indicate, Danchev's method of measuring rock color cannot be used, especially in photometric measurements from the air.

In 1954, Vistelius and Yaroslavskaya published their observations on the objective evaluation of the chromaticity of sandy-silt terrigenous Cretaceous deposits in the Transcaspia. The authors obtained measurements of the color characteristics by photometry of the rocks on the FM-2 universal photometer with a subsequent conversion to chromaticity coordinates in the international colorimetric XYZ system. The authors converted the spectral luminance factors of the measured specimen to chromaticity coordinates (x, y, and z) by the procedure described by Gurevich (1950). Vistelius and Yaroslavskaya succeeded in establishing a definite relationship between the chromaticity coordinates of terrigenous Cretaceous deposits and the conditions of their accumulation and showed that deposits of each terrigenous mineralogical province had their own characteristic chomaticity coordinates. The authors also found a relationship between the chromaticity coordinates x, y, and z (Vistelius and Yaroslavskaya, 1954).

These authors did not discuss the factors determining or influencing the color characteristics of the investigated rocks.

In 1954-1955 the author carried out similar work on the measurement of the color characteristics of sandy-silt deposits of the red beds of Cheleken Peninsula. The methods of measurement and the calculation procedure were the same as those used by Vistelius and Yaroslavskaya, but the investigations were supplemented with quantitative mineralogical calculations of the composition of the rocks and chemical analyses to determine their content of soluble ferrous and ferric iron. These investigations showed that the content of hematite fragments and soluble ferric iron affected the hue of the rock. The effect of the latter was so strong that a very slight increase in the Fe_2O_3 content of the rock shifted the chromaticity coordinate of the rock into the red region of the spectrum and increased the purity of the color (Romanova, 1958)

The above examples of the objective evaluation of color characteristics of rocks show:

1) the unwieldiness of the method of obtaining color characteristics of an investigated rock when the primary measurements made by photometry of the sample are then converted by means of fairly complex formulas to obtain the color parameters of the rock;

2) that there are an infinite number of groups of three mutually independent colors from which a given color can be obtained by mixing (from the laws of mixing of colors).

All the above points regarding the color characteristics of rocks led us to select the reflecting properties of rocks for aeropetrographic mapping. We were assisted here by the progress in the field of optical science. For instance, the development of methods of constructing diffraction gratings (Gerasimov et al., 1958) has made possible the construction of spectrometers with high resolution and the investigation of reflection spectra in a wide spectral region — from the ultraviolet to the near infrared. There has been a parellel development of methods of recording reflection spectra of rocks from the air — from an airplane or helicopter — and of methods of evaluating spectral luminance curves.

A logical outcome of the above observations was the idea of evaluating the curves of the spectral luminance factors and the geological interpretation of these curves. Thus, we directed all our subsequent observations toward a study of the reflecting properties of rocks, measurement of their spectral luminance factors, and a numerical evaluation of the obtained curves of spectral luminance factors.

To simplify the problem we chose rocks with a surface reflecting incident radiation diffusely, or orthotropically, as the first objects of investigation. Objects of such a kind are blown sands, devoid of plant or soil cover. Since blown sands comprise extensive areas of existing sandy deserts, the investigation of their reflecting properties could be combined with the tackling of several geological problems connected with the investigation of the material composition of the deposits in these deserts.

History of research on luminance characteristics of natural objects. The idea of the objective evaluation of the luminance characteristics of natural objects is by no means a new one. Hipparchus foresaw the progress of photometry in freeing the human eye from the estimation of brightness equalities. A thousand years later, Fersman (1936) expressed his opinion on the importance of the new method of scientific analysis of rocks when "the spectroscope and camera are combined with the eye...."

Experimental observations of the reflecting properties of natural objects have been the subject of numerous works in the field of aerial photography. This has involved a rational choice of photographic materials and conditions of photography.

A study of the reflecting properties of natural objects on earth on a large scale is being conducted in the field of planetary astronomy so that the luminance of terrestrial objects can be compared with the luminance of surfaces of other planets, and there is a vast specialist literature on this subject.

The reflecting properties of terrestrial objects are being studied for practical purposes by specialists in quite different fields. The results of measurements of the reflecting characteristics of natural objects in connection with specific problems in the fields of illumination engineering, health resort studies, soil science, forestry, botany, mineralogy, and petrography have been published.

In the twenties of this century, measurements of the reflection factors of sand, loam, herbage, sawdust, and freshly fallen snow were carried out for the purpose of determining the best location of lighting points. The reflection factors were measured with an Ulbricht photometer. The results of the measurements showed that the reflection of light energy from a surface follows a very complex law and that the reflection factor depends on the angle of incidence of the light beam, the state of the ground cover, and the nature of the investigated cover (Kulebakin, 1926).

The reflecting properties of some natural objects and the spectral albedo of the earth's surface were investigated by Kalitin (1929, 1931, 1938) with the aim of maximum utilization of solar radiant energy as a health factor. He designed a special instrument — a pyranometer, based on the same principle as the Ångström pyranometer (1919) but adapted for measurement of reflected radiant energy.

In the thirties, investigations on the reflecting properties of plants, soils, and other terrestrial objects were begun. This research was conducted on the grounds of the Pulkova Observatory under the supervision of G.A. Tikhov. The first measurements of reflecting properties were made for the purpose of solving some problems of aerial photography. Photographic photometry was used in this case. This method was later employed in connection with a study of the reflecting properties of natural objects on other planets with particular reference to the development of astrobotany (Tikhov, 1950).

Several papers dealing with investigations of the spectral reflectance of terrestrial formations and methods of investigating it have been published in special collections edited by G.A. Tikhov (Sharonov, 1934; Krinov, 1934, 1935).

Considerable attention has been and is still being devoted to the investigation of the reflecting properties of vegetation. Atlases of spectral luminance curves of vegetation from different climatic zones have been compiled (Kozlova, 1955). In measurements of the reflecting properties of flowers, Tikhov discovered the additional emission in the infrared by many kinds of flowers (Tikhov, 1948). A similar effect is known in the literature as the "phenomenon of Elisabeth Linnaeus," but the results of the observations have not been measured (Minnaert, 1958, p. 174).

In further studies of the reflecting properties of vegetation and timber, spectrally zoned films and appropriate light filters were used. This greatly improved the identification of characteristics of forest vegetation on aerial photographs (Artsybashev, 1957, 1959; Belov, 1959; Belov and Berezin, 1958).

The collection of information on the reflecting properties of various natural objects (vegetation, geological objects, snow, water surfaces, etc.) enabled Krinov to compile a special summary. In his summary Krinov suggested an original spectrophotometric classification of natural objects. All geological formations were united in one class (Krinov, 1947).

The idea of using a spectroscope in conjunction with a camera for the photometry of natural objects from the air originated approximately half a century ago. In 1914 N.A. Morozov obtained several spectrograms of the earth's surface from a balloon. In 1929 V.A. Faas used a spectroscope to obtain spectrograms of natural formations from an airplane. The spectrograms obtained in this case were, unfortunately, unsuitable for analysis, and the premature death of the investigator prevented him from putting the method to practical test.

The cited data show that the use of the values of the spectral luminance of rocks for determining their type composition will permit the solution of several geological and petrographic problems.

The development of the method of determining the composition of rocks from the air by their reflecting characteristics necessitated the conduction of a number of experiments and the solution of a series of problems. An account of the results is given in subsequent chapters.

CHAPTER II

BASIC PHOTOMETRIC CONCEPTS

In the subsequent account we will constantly have to deal with certain photometric quantities and other concepts which are not a familiar part of the geologist's work. The concepts which we will have to discuss include the scattering (diffusion) of radiant energy by reflection from the surface of an object, the types of indicatrices of diffusion, the luminance of a reflecting surface, the spectral luminance, and the spectral luminance factor. Definitions of these concepts and their evaluation can be found in several textbooks, particularly those of Landsberg (1957), Fabry (1934), Tikhov (1950), Volosov and Tsivkin (1960), and others.

The chapter concludes with a descriptive account of experimental measurements of the spectral luminance factors of rock samples. The aim of these measurements was to select geological objects for spectrophotometry from the air.

Reflection and Indicatrices of Diffusion

We distinguish any visually considered object by the way in which it emits light or reflects light energy fallin on it.

Rocks and minerals are non-self-luminous bodies and they can be distinguished visually from one another by their color, luster, and degree of transparency, which in toto constitute the reflecting properties of the considered rock or mineral sample. The eye registers only the small part of the radiant energy within the visible region of the spectrum (400-680 mμ).

If we replace the eye by some other detector of reflected radiant energy, we can enlarge the wavelength region accessible to investigation and determine the nature of the reflection not only of light energy, but also of radiant energy in the short-wave (ultraviolet) or long-wave (infrared) regions of the spectrum according to the properties of the registering device.

In the first chapter we showed that the reflecting properties of rocks can be objectively measured much more easily and accurately from the luminance properties than from the color characteristics. In view of this we must discuss some of the general laws of reflection of radiant energy.

The scattering of incident radiant energy by any reflecting surface depends on the properties of the surface. Part of the incident energy is spent on heating the body, i.e., it is absorbed by the body; part of the flux,

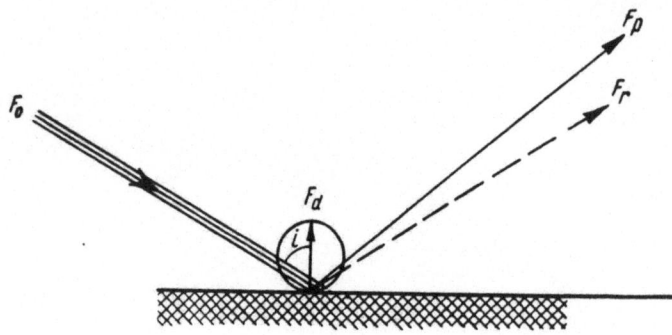

Fig. 2. Scattering of incident light flux F_0 by a reflecting surface. F_p, the total reflection in a given direction, is made up of the sum of the directly reflected flux F_r and the diffusely reflected flux F_d.

depending on the degree of transparency of the object, may be transmitted; and part of the incident radiant flux is reflected by the surface of the body.

When a rock is observed from a helicopter or an airplane only the reflected part of the radiant energy can be registered.

The properties and laws of behavior of a light beam passing through a rock or a particular mineral lie in the field of optics and crystal optics and are the main properties used for the identification of minerals in microscopic petrography and mineralogy.

The properties and laws of behavior of a beam of light or energy reflected by the surface of a rock underlie the method expounded in this book, and in view of this we must analyze some very simple cases of reflection.

In the general case in which a radiant flux F_0 falls at an angle i onto the surface of a rock (Fig. 2), the total radiant flux F_p reflected from this scattering surface is made up of two fluxes: the scattered or diffusely reflected flux F_d and the directly or specularly reflected flux F_r. The total reflection factor ρ of the investigated surface is equal to the ratio of the sum of these two fluxes to the incident radiant flux,

$$\varrho = \frac{F_d + F_r}{F_0} \, .$$

The distribution of the reflected radiant energy will vary for different objects; it will depend on the nature of the investigated surface, the angle of incidence of the light flux, and the composition and properties of the object.

The reflecting properties of a scattering surface are usually represented graphically in the form of polar diagrams constructed from the radius vectors corresponding to the amounts of energy reflected in each direction. Such diagrams are called indicatrices of diffusion (see Figs. 3-8). The nature of the indicatrix of diffusion for most objects depends on the angle of incidence i of the light flux and on the angle ε at which the instrument for registering the reflected light flux is situated.

The indicatrix of diffusion of an investigated surface is plotted by measuring ρ at particular angles of incidence of the light flux. Practice has shown that it is sufficient to make measurements for angles i equal to 10, 20, 30, 40, 50, 60, 70, and 80°. If the angle of registration of the radiant flux is altered, which can easily be done by rotation of the photometer, then several diagrams corresponding to particular angles ε can be constructed. Examples of the construction of such indicatrices of diffusion obtained by measurement of several rock samples on the large indicatometer of Leningrad State University are shown in Figs. 3-8. The arrows on the figures indicate the position of the photometer for all the angles of incidence (indicated by the radii) of the light flux.

From the nature of the indicatrix of diffusion, we can distinguish surfaces which give diffuse (or orthotropic) reflection, or mirror, glasslike, metallic, and mixed types of reflection.

Diffuse reflection is observed when radiation falls on an orthotropic surface. In this case the reflected beam of radiant energy is scattered equal in all directions, irrespective of the angle of incidence of the radiant flux. The indicatrix of diffusion of an ideally reflecting surface approximates a sphere. The reflection factor in this case will be equal to the ratio of the diffusely reflected flux F_d to the incident flux F_0.

$$\varrho = \frac{F_d}{F_0} \, .$$

Among geological objects, the surfaces of blown sand are diffusely reflecting surfaces. The nature of the indicatrices of diffusion of different types of sands is shown in Figs. 3-5. The sands for investigation were taken from barchans in the central part of the Kara-kum Desert, near Darvaza (Fig. 3), from the Central Kyzyl-kum (Fig. 4), and from the region of the Sulak delta (Fig. 5). As the figures show, in the case of photometry at angles not exceeding 50°, the indicatrices of diffusion of the luminance factors are spherical; i.e., the surfaces of the sands reflect the light energy orthotropically (diffusely).

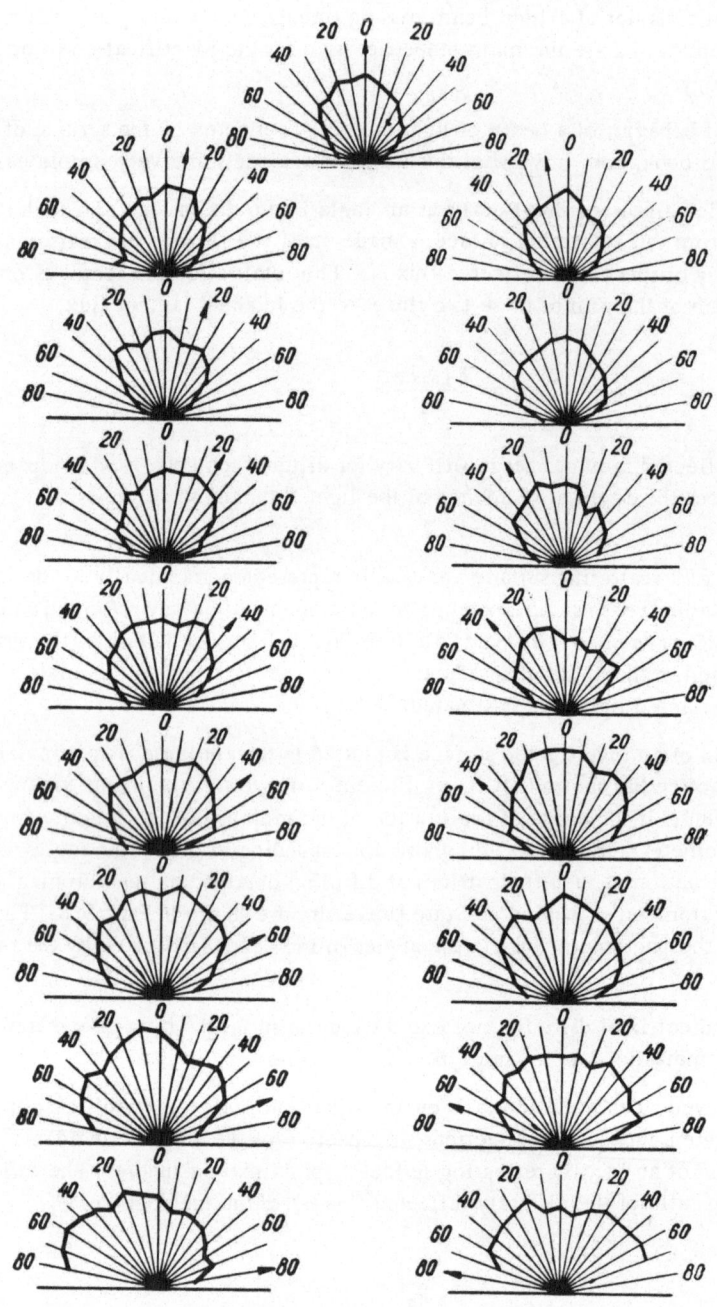

Fig. 3. Indicatrices of diffusion of sand from a barchan in the Darvaza region (Central Kara-kum).

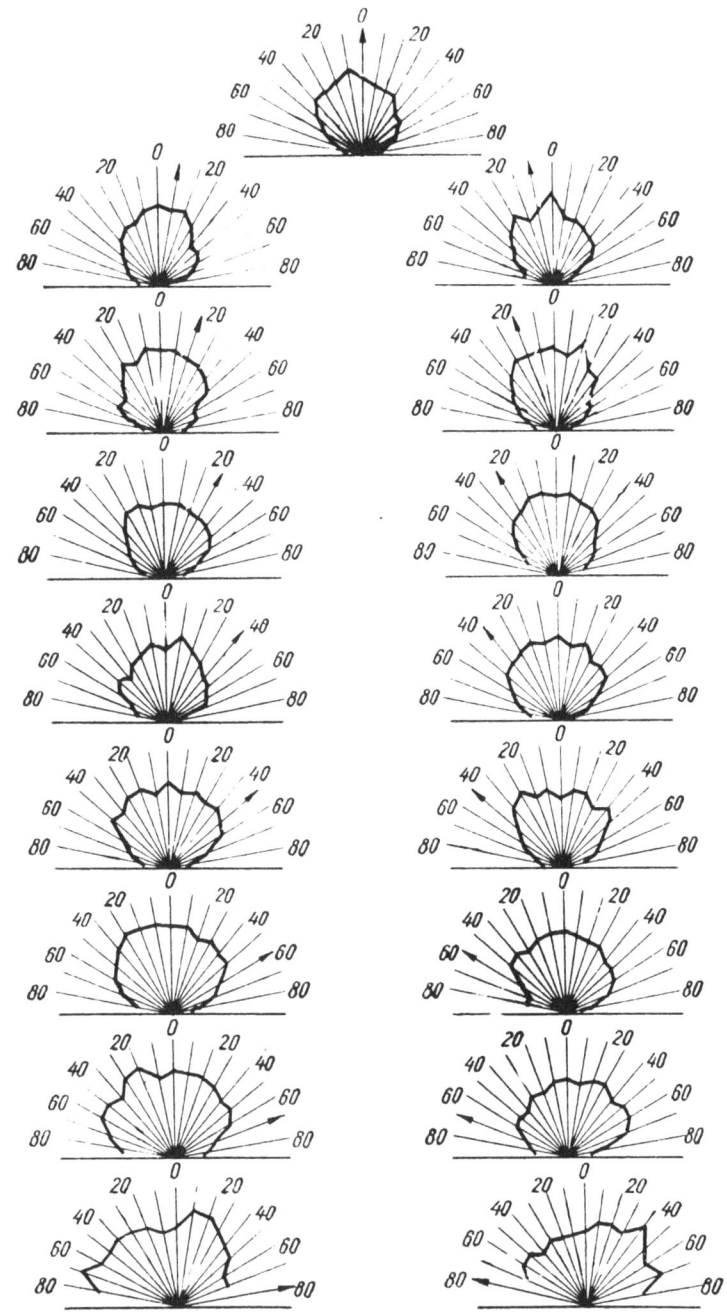

Fig. 4. Indicatrices of diffusion of sand from a barchan in the
Tamdy-Bulak region (Central Kyzyl-kum).

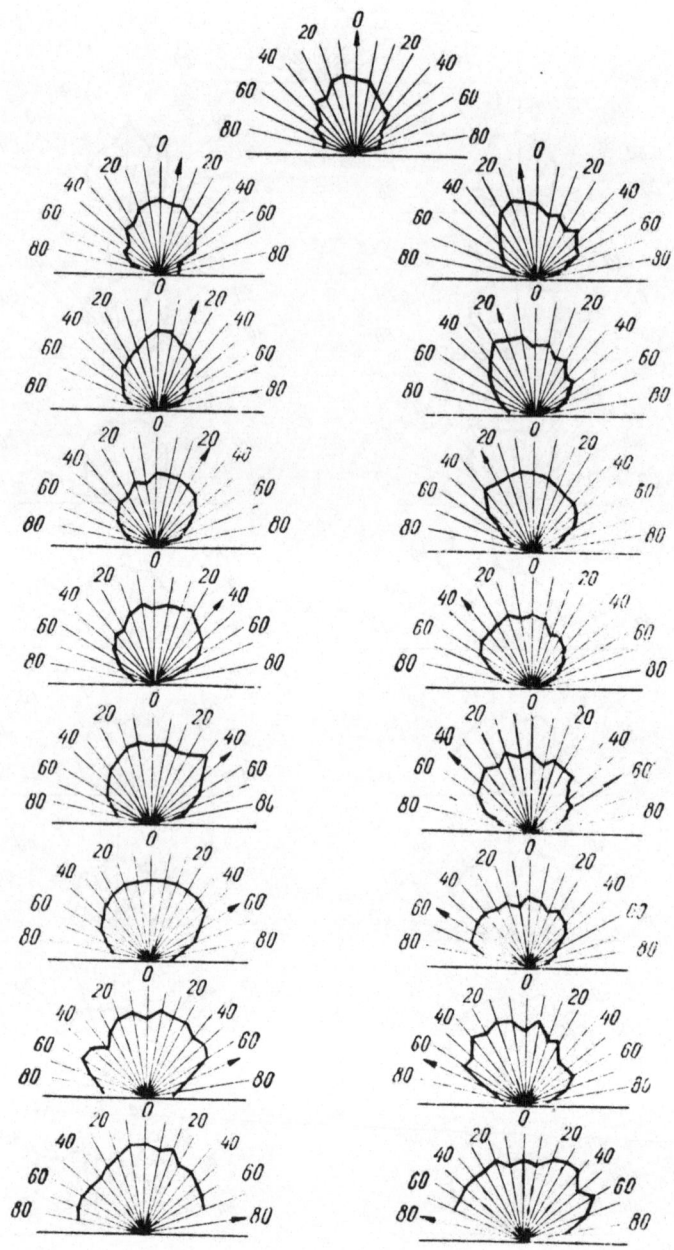

Fig. 5. Indicatrices of diffusion of sand from a barchan in the region of the Sulak delta.

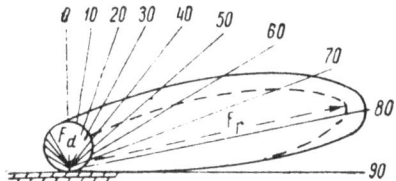

Fig. 6. Indicatrix of light strength of ashpalt coatings (according to M.A. Ostrovskii). F_d) Diffusely reflected light flux; F_r) directly reflected light flux.

Specular reflection is observed when a beam of light falls on an optically polished surface, which has only small irregularities in comparison with the wavelength. In this case the incident beam of rays is reflected in accordance with the well-known law of reflection (angle of incidence equal to angle of reflection), and the reflection factor is equal to the ratio of the specularly reflected flux F_r to the incident flux F_0,

$$\varrho = \frac{F_r}{F_0} .$$

Among natural objects, smooth water surfaces exhibit specular reflection. Well-rolled asphalt roads (Fig. 6) are to some extent specularly reflecting surfaces. Among geological objects, specular reflection is given by deposits of crystalline salts and solid outcrops of igneous rocks, especially if they have joints or oriented prototectonics.

Glasslike reflection occurs in the case where the reflected surface is transparent and absorbs hardly any of the radiant flux incident on it. Here the incident light flux is almost entirely transmitted by the medium, and the investigation of the properties of the medium involves measurements of the refractive indices. Among natural objects, glasslike reflection is given by an ice surface, a frozen snow crust, and certain minerals.

Metallic reflection is characteristic of strongly absorbing media. Most metals give this type of reflection.

Mixed reflection is the type which occurs most frequently in nature. In this case the reflecting surface scatters the incident radiant energy diffusely and directly (specularly). The indicatrix of diffusion in this type of reflection is made up of two indicatrices – the spherical F_d indicatrix and the elongate-ellipsoidal F_r indicatrix. Depending on which of these predominates we obtain an indicatrix of diffusion similar to that of a diffusely reflecting surface or a surface giving specular reflection. For instance, in the case of reflection from asphalted roads (Fig. 6) the indicatrix of reflection is made up of a noticeably predominant specular reflection and a small diffuse contribution. The nature of the reflection of most geological objects is of the mixed type. An example of this is provided by the indicatrices of reflection from a medium-grained biotite granite (Fig. 7) and from porphyrite (Fig. 8).

The granite surface was the natural cleavage face obtained when a granite sample was broken off from outcropping rock on the

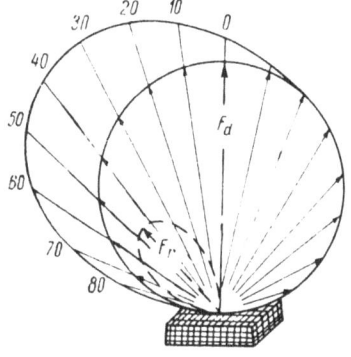

Fig. 7. Indicatrix of luminance factor of medium-grained biotite granite from Mt. Shakh-Adam (Krasnovodsk Peninsula). F_d) Diffusely reflected light flux; F_r) directly reflected light flux.

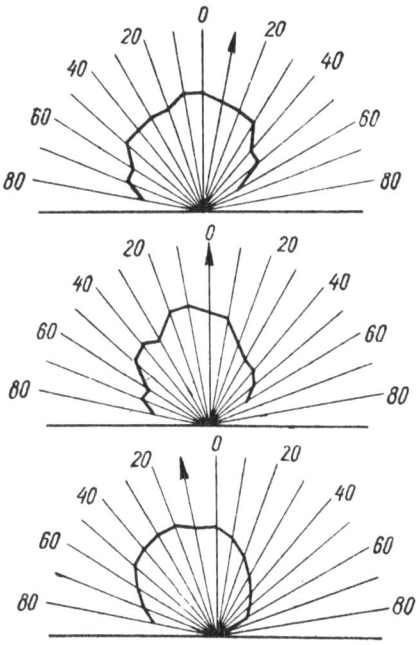

Fig. 8. Indicatrix of diffusion of pyroxene porphyrite from the Ufra Peninsula (Caspian Sea).

slope of Mt. Shakh-Adam (Krasnovodsk Peninsula). The indicatrix in this case (Fig. 7) is made up of two indicatrices of diffusion, but with an appreciable predominance of the diffusely reflected light flux F_d over the specularly reflected F_r.

The indicatrices of diffusion obtained for a dark, dense pyroxene porphyrite from the Ufra Peninsula (Caspian Sea) by the same method as for the granite are shown in Fig. 8. The measured face was the natural surface of the slightly weathered porphyrite, split off along the direction of a joint. The indicatrices of diffusion of such a surface outwardly resemble the indicatrices of diffusely reflecting bodies. In any case, in the obtained indicatrices of diffusion, diffuse reflection predominates appreciably over direct or specular reflection.

The indicatrices of luminance of herbage and other natural objects for which there is published information (Krasil'shchikov and Pyatkovskaya, 1957; Orlova, 1958) are, in the vast majority of cases, representative of the mixed type of reflection.

Albedo, Luminance, Spectral Luminance

The part of the radiant energy reflected from the surface of any object can be measured by some method or other. The total reflection, i.e., the whole radiant energy reflected by a given surface, cannot be measured in every case, since this requires special and rather cumbersome devices. One of these devices – the Ulbricht sphere – is widely used in laboratory practice. In measurement of the total reflection of light energy with this spherical photometer the illumination of the internal surface of the sphere due to reflection from the surface of the test object is compared with the illumination of the same surface due to reflection from some standard.

If the whole flux F reflected by the given surface in all directions is measured and its ratio to the total flux F_0 incident on the test surface is calculated, this ratio is generally called the albedo A, i.e., the whiteness,

$$A = \frac{F}{F_0} \, .$$

Usually we can measure only a small part of the reflected radiant flux, bounded by some small solid angle in a given direction $d\Omega$. The emission of the illuminated surface in a given direction determined by the angle ε with the normal to this surface is called the luminance of the reflecting surface or the surface luminance B_ε. If a radiant flux Φ falls on the surface of a non-self-luminous body (Fig. 9), the luminance B_ε of this surface is determined from the following relationship:

$$B_\varepsilon = \frac{d\Phi}{\sigma \, d\Omega \cos \varepsilon} \, ,$$

where $d\Phi$ is the part of the radiant flux reflected from the investigated area; σ is the area from which the reflection occurs; $d\Omega$ is the small solid angle within which the reflected light flux is measured and is equal to the ratio of the area which the angle cuts out on the surface of a sphere with its center at the apex of the angle to the square of the radius of this sphere; ε is the angle between the given direction and the normal to the reflecting surface.

The standard unit of luminance is equal to the luminance of the minimum plane surface, radiating equally in all directions, such that the ratio of the light intensity in candles to its area in square meters is equal to unity. This unit is called the nit (nt). A nonstandard unit of luminance of an illuminated surface is the apostilb (asb), or lux on a white surface, i.e., the luminance

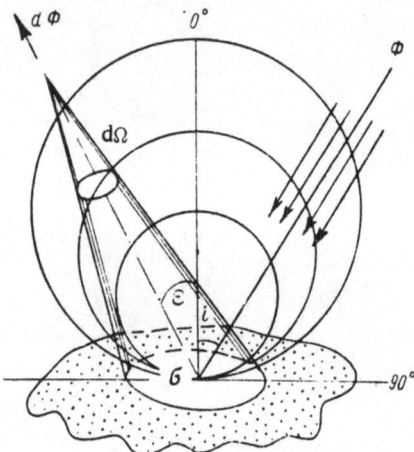

Fig. 9. Diagram illustrating the measurement of luminance of a surface as distinct from the albedo. Φ) Incident radiant flux; $d\Phi$) reflected radiant flux measured within a solid angle $d\Omega$; σ) area from which reflected radiant flux $d\Phi$ is measured; ε) angle between given direction and normal to reflecting surface; i) angle between direction of incident radiant flux and normal to reflecting surface.

of an ideal diffuser on which an illuminance of one lux is created; the luminance is measured in a direction perpendicular to this surface.

The spectral luminance B_λ is the luminance of a reflecting surface measured at a particular wavelength λ.

The ratio of the spectral luminance of an investigated surface to the spectral luminance of a standard in identical conditions of illumination is called the s p e c t r a l l u m i n a n c e f a c t o r ρ_λ.

Special instruments, called spectrometers, are used for measurement of the spectral luminance. Almost all the existing spectrometers are designed for measurement of the relative spectral luminance. The reliability of the obtained measurements primarily depends on the properties of the chosen standard and on the nature of its indicatrix of diffusion. In view of this, we must make an assessment of the most frequently used standards.

Standards. As already stated, the spectral luminance is measured as the ratio of the luminance of the investigated object at a given wavelength to the luminance of a standard (at the same wavelength). In view of this, the question of standards is of very great importance. In practical work any surface which approaches an ideal diffuser in the nature of its reflection can be taken as a standard. An ideal diffuser is a surface which completely reflects all the radiant energy incident on it and scatters it uniformly in all directions irrespective of the wavelength of the incident rays. The reflection factor of an ideal diffuser is equal to unity for all directions of incidence of the radiant flux and for all wavelengths. The reflecting properties of a mat white surface approximate those of an ideal diffuser.

Thus, the surface adopted as a standard must satisfy two requirements: 1) preservation of orthotropy, i.e., the indicatrix of diffusion of this surface must in all cases be close to a sphere, irrespective of the angle of incidence of the radiant flux (angle i); 2) preservation of neutrality, i.e., the nature of the spectral luminance curves must be independent of the spectral composition of the incident energy.

Plates smoked with magnesium oxide, which reflects incident energy orthotropically and neutrally, represent the closest approach to an ideal diffuser. They have been widely used in experimental practice. Numerous papers have been devoted to a description of the properties of these plates (Abishev, 1948; Benford, Lloyd, Schwarz, 1948; Sanders, Middleton, 1953; Waldron, Tellex, 1955).

Despite the high reflecting qualities of plates coated with magnesium oxide, their use is unfortunately limited mainly to work in stationary or laboratory conditions. The reason for this is the complexity of the process of depositing a uniform coating and the difficulty of preserving the purity of the obtained surface in field conditions. In view of this, barite paper is usually used in field work. Ostwald suggested that a surface of chemically pure barium sulfate could be used as a standard if the barium sulfate layer was sufficiently thick. The standard consisted of a plate painted with zinc white and then coated with five layers of barium sulfate and polished. It was later discovered that the thickness of the barium sulfate layer could be reduced if thick white paper was used as a backing. Thus, barite paper has been widely adopted as a standard reflector for the photometry of various natural objects.

The reflecting properties of the barite paper used as a standard in our investigations are shown in the form of indicatrices of diffusion of some grades (Figs. 10-13) and in the form of the spectral characteristic of one of the paper samples (Fig. 14).

The indicatrices of diffusion obtained for different grades of barite paper by measurement on the indicatometer of the Leningrad State University showed that the reflecting properties of barite paper in many cases do not correspond to those of an ideal diffuser, since barite paper frequently fails to give orthotropic reflection. When the light flux has angles of incidence of more than 45°, the specular component becomes appreciable in the scattering of light reflected by barite paper and reaches a maximum when the angle of incidence of the light flux is in the range 70-80°.

The neutrality of the barite paper which we used as a standard is sufficiently well maintained. At any rate, the required neutrality of reflection is preserved within the visible region of the spectrum and in the near-infrared (see Fig. 14).

When the indicatrices of diffusion of barite paper were measured through different color filters, the obtained indicatrices were of the same general nature. The types of indicatrices of diffusion obtained for barite paper by measurement with green and red filters (see Figs. 12, 13) showed that they were approximately the same as those obtained in measurement without a filter.

A barite-paper standard is fairly durable in work. It can be kept in special holders in the field and if the surface becomes dirty the used sample of barite paper can easily be replaced by a new piece of the same paper. The photometry of a barite-paper standard mounted in a holder is illustrated in Fig. 15. The photometry is performed on the ground, in an area free from any shadows and additional illumination.

During a flight in an airplane or helicopter it is difficult to carry out the photometry of barite paper, and hence several types of standardizing devices have been proposed in place of barite paper. The fitting of such standardizing devices aboard the airplane or helicopter dispenses with the need for extra landings and ensures the minimum lapse of time between photometry of the standard and photometry of the investigated object. As an example we describe the standardizing device developed and adopted in practical field photometry by a group of workers in the Laboratory of Aeromethods, Academy of Sciences, USSR (Voronkova et al., 1960). The standardizing device is a plane, milk-white glass system mounted in the upper end of a long tube. The lower end of this tube can be fitted over the objective of the photometer (see Fig. 50). The tube of the standardizing device is rigidly attached to the frame which holds the spectrophotometer in its working position on board the aircraft. The inside surface of the tube is painted mat white. Thus, the radiant flux passes through the flat mat glass, traverses the length of the tube, where it is reflected repeatedly from the white walls, and falls on the objective of the photometer.

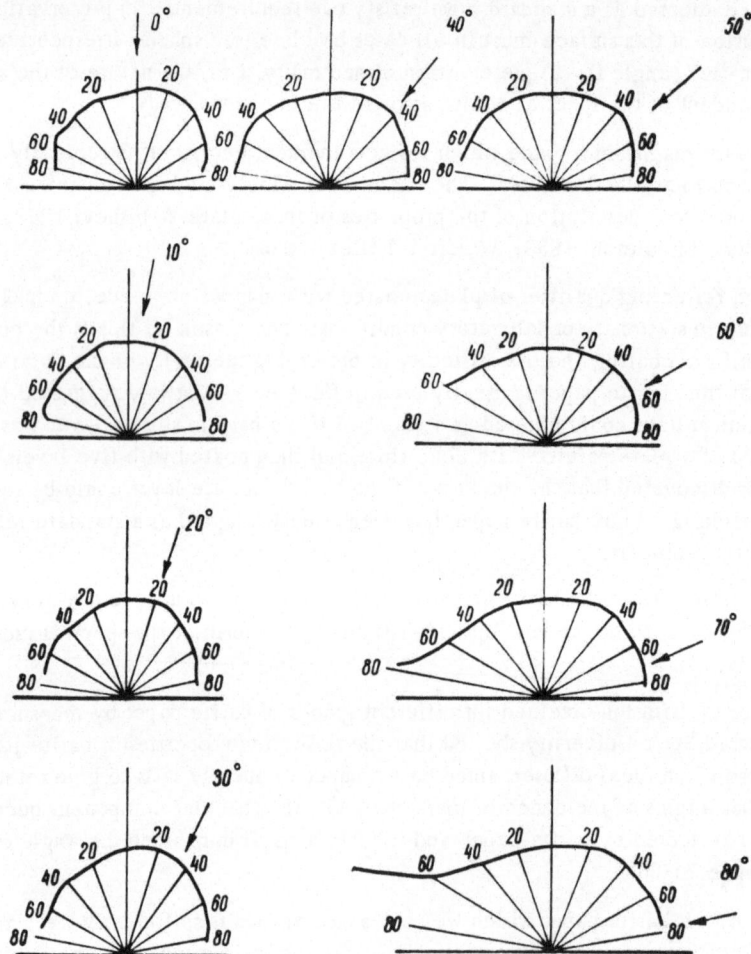

Fig. 10. Indicatrices of diffusion of barite paper for angles of incidence of the light flux (arrows with degrees) and angles of registration of the reflected light flux in the range 0 to 80° (according to N.V. Eliseeva).

16

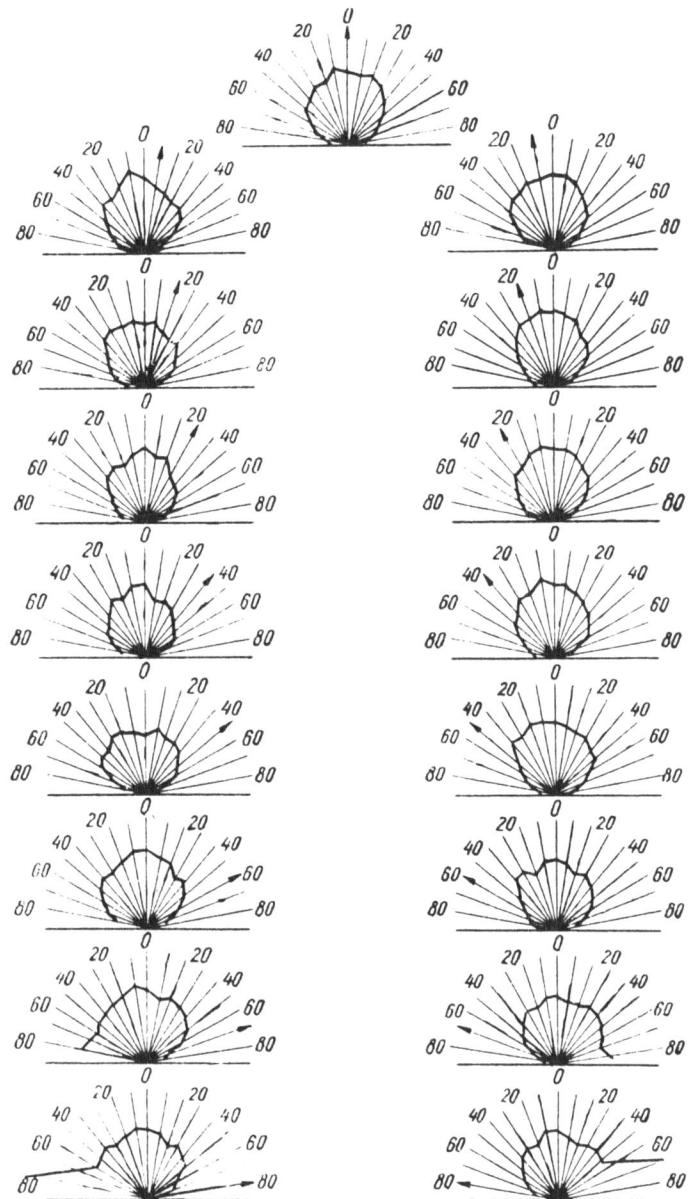

Fig. 11. Indicatrices of diffusion obtained for a 1960 sample of barite paper by measurement on the indicatometer of Leningrad State University. The arrows show the position of the photometer for different angles of incidence of the light flux.

Fig. 12. Indicatrices of diffusion of a 1960 sample of barite paper. The
measurements of ρ were made with a green filter.

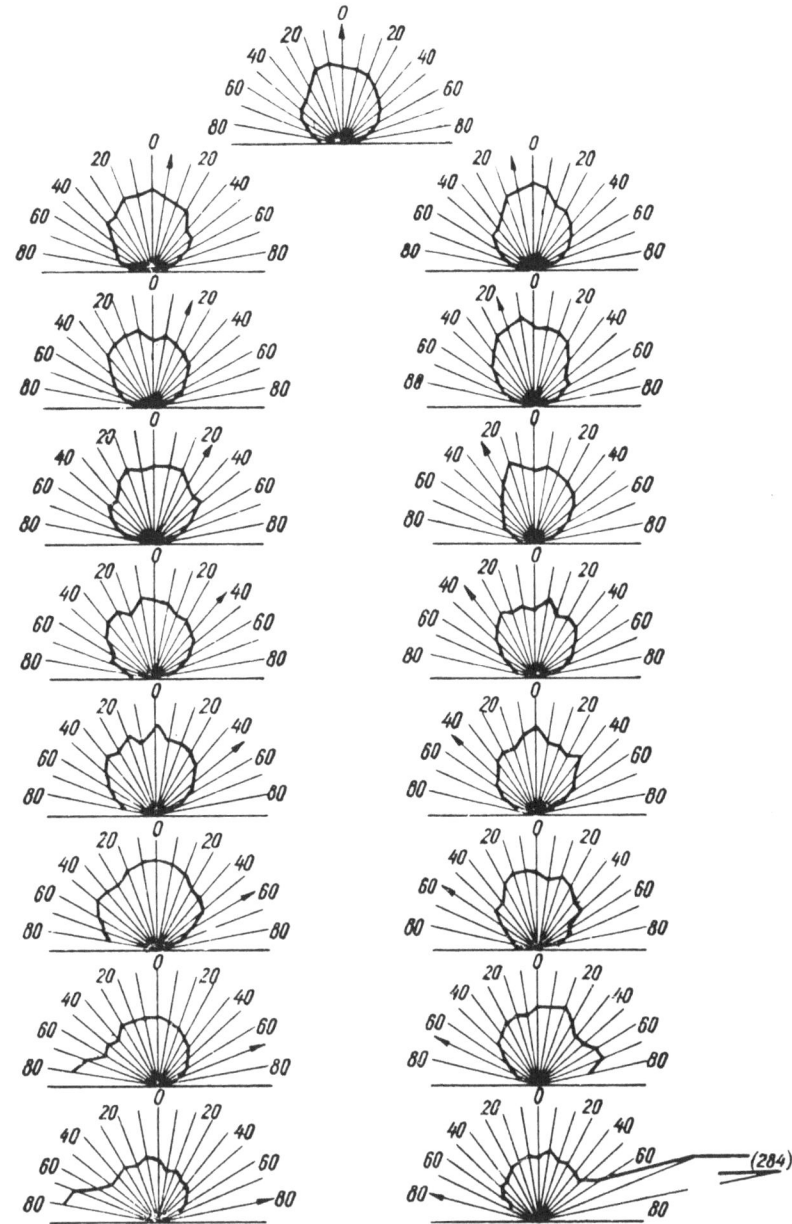

Fig. 13. Indicatrices of diffusion of a 1960 sample of barite paper. The
measurements of ρ were made with a red filter.

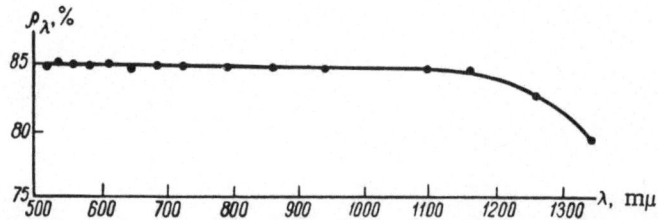

Fig. 14. Spectral reflection of barite paper used as a standard
in our work (data of L.B. Krasil'shchikov et al., 1958).

The standardizing device is fitted in such a way to the frame holding the spectrophotometer that the photometer can be rotated around a horizontal axis. This allows successive photometry of the investigated object and the standard at relatively short intervals. The shortening of the period between the times of photometry of the standard and the object reduces the error due to variation in light intensity at these times.

For calibration of the described standardizing device the data of its photometry must be compared with the data of the photometry of a general standard – a plate smoked with magnesium oxide, or barite paper. In this case photometry of the standardizing device and the standard should be carried out in absolutely indentical conditions of illumination.

Fig. 15. Photometry of standard (barite paper) when the helicopter lands. Photograph by K.E. Meleshko.

Spectral Luminance Curves and Their Evaluation

The measured values of the spectral luminance factors of investigated objects are usually represented as curves in the system of coordinates wavelength vs. spectral luminance factor. The wavelengths in millimicrons or angstroms are plotted on the x axis, and the values of the measured spectral luminance factors ρ_λ as a percentage relative to the chosen standard are plotted on the y axis. The measured ρ_λ given in Table 1 are represented graphically in Fig. 16. The rock samples listed in Table 1 were measured on the FM-2 universal photometer with a set of interference filters.

The spectral luminance curves plotted for various geobotanical objects from the data of Belov and Artsybashev (1957) are shown in Fig. 17. The types of spectral luminance curves for various objects in accordance with Krinov's spectrophotometric classification are shown in Fig. 18 (Krinov, 1947). These figures show that the types of spectral luminance curves for natural objects are quite different as regards shape and general position of the curve on the coordinate plane (ρ_λ, λ) for the different formations.

Experimental Measurements of Spectral Luminance Factors of Rock Samples

To discover the factors affecting the values of ρ_λ of the investigated object and associated with the photometry conditions, we carried out several experimental measurements of ρ_λ. In each case we altered one of the conditions affecting the value of ρ_λ and kept the rest constant.

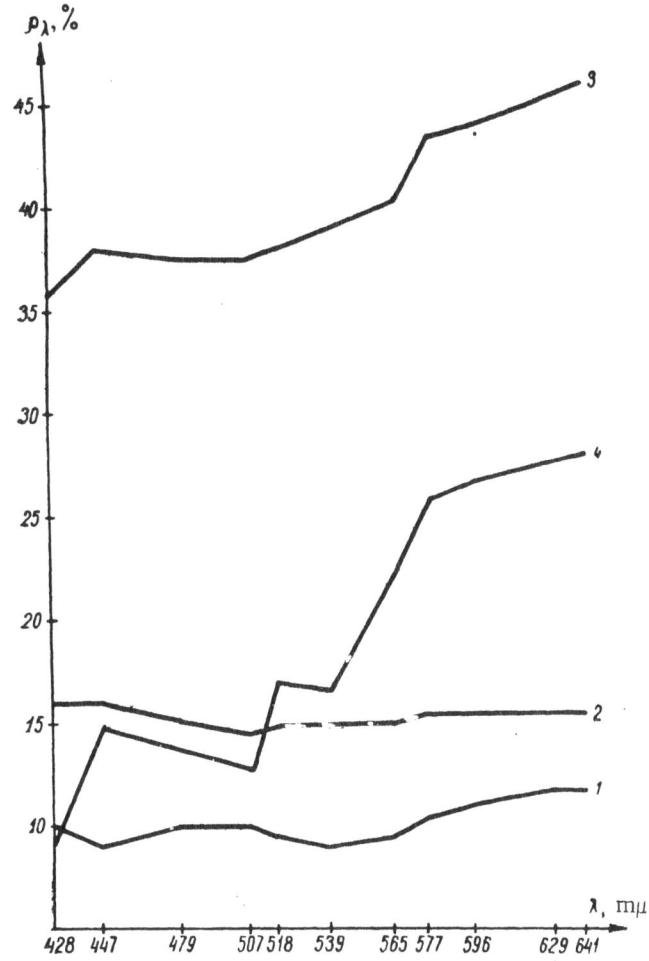

Fig. 16. Spectral luminance curves obtained for some rock samples by measurement on the FM-2 universal photometer (the data of the measurements are given in Table 1): 1) Pyroxene porphyrite from Ufra Peninsula 2) amphibole gabbro from Krasnovodsk Peninsula; 3) biotite granite from Mt. Shakh-Adam; 4) graywacke medium-grained sand from a barchan near Darvaza.

It is known from the general principles of photometry that the values of ρ_λ for an investigated object can be affected by the conditions of illumination of the measured surface and the standard with which the spectral luminances of the investigated object are compared, the nature of the measured surface of the investigated object, the properties and mineralogical composition of the investigated rock, and the photometric properties of the standard.

Effect of Natural Illumination on Values of ρ_λ of Rock

The conditions of illumination affecting the values of ρ_λ of the investigated object (rock) include the angle of incidence of the light (or radiant) flux on the photometered surface, the nature of the natural illumination, the presence of clouds, and the effect of dust, haze and mist.

Effect of angle of incidence i of the light flux on values of ρ_λ. The effect of the angle of incidence i of the light flux on the values of ρ_λ was determined by repeated photometry of the same rock sample at different times of the day. The photometry was performed with the FM-2 universal photometer and a field spectrometer.

TABLE 1. Data of Photometry of Some Rock Samples

Wave-length, mμ	Spectral luminance factors, %			
	porphyrite	gabbro	granite	sand
428	10.0	16.0	36.0	9.0
447	9.0	16.0	38.0	14.7
479	10.0	15.0	37.5	13.8
507	10.0	14.5	37.5	13.0
518	9.5	15.0	38.0	17.0
539	9.0	15.0	39.0	16.7
565	9.5	15.0	40.5	22.2
577	10.5	15.5	43.5	25.7
596	11.0	15.5	44.0	26.8
629	11.5	15.5	45.5	27.7
641	11.5	15.5	46.0	28.0

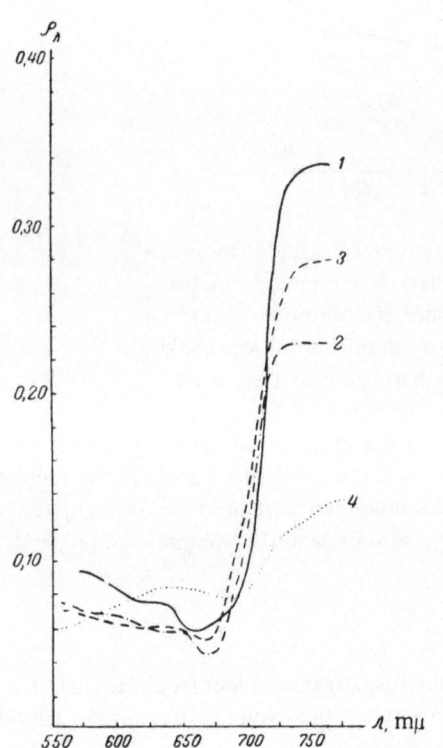

Fig. 17. Spectral luminance curves for various geobotanical objects (from Belov and Artsybashev): 1) Dry meadow; 2) reforested clearing; 3) non-reforested clearing; 4) peat bogs.

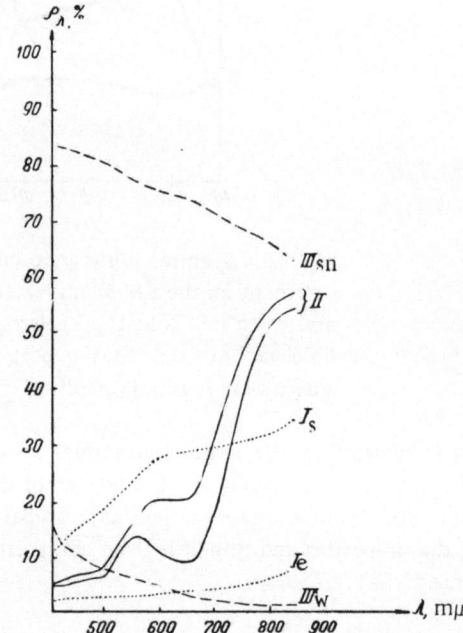

Fig. 18. Classification of spectral luminance curves of natural objects (according to Krinov): I_e) Curves of soils, dirt roads, $\gamma = 1.64$; I_s) curves of sands and rocks, $\gamma = 2.71$; II) curves of tree plantations in summer and autumn, $\gamma = 2.45$-2.56; III_w) curves of water surface, $\gamma = 0.19$; III_{sn}) curves of freshly-fallen snow, $\gamma = 0.88$. According to Krinov, γ is the ratio of the values ρ_{650}:ρ_{400} mμ.

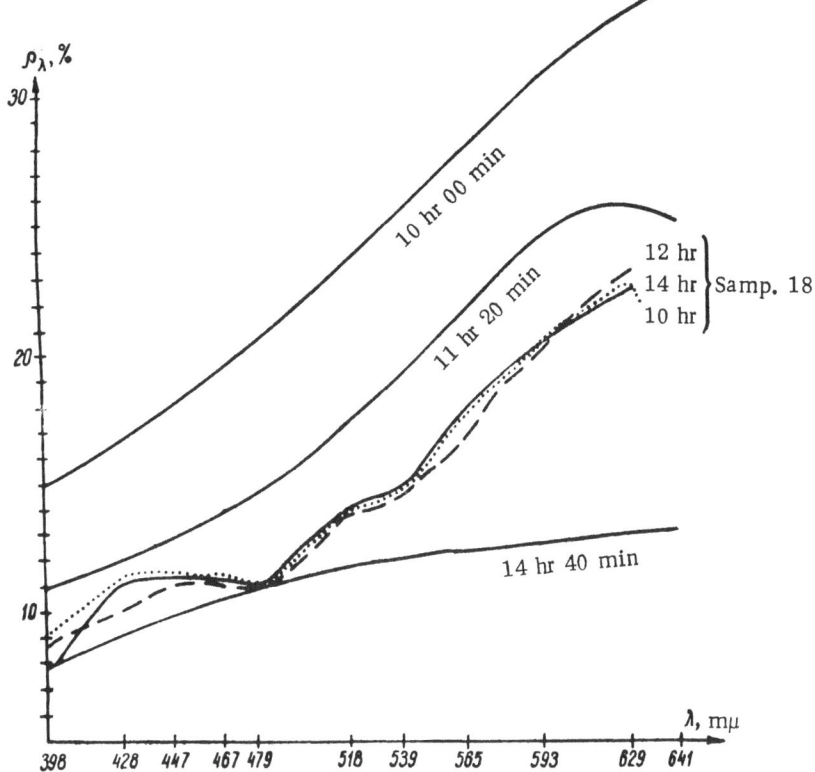

Fig. 19. Spectral luminance curves obtained for blown sand (sample 18) and cemented carbonate sandstone at different times of day. The photometry was performed with the FM-2 universal photometer in Nebit-Dag on December 18, 1956.

During the measurements the following conditions were observed:

1) the same part of the rock outcrop or rock sample was photometered;
2) the measurements were made by the same observer;
3) photometry was conducted under a completely clear, cloudless sky;
4) the positions of the instrument and objects were the same in relation to the points of the compass.

Thus, if photometry is conducted at different times of the day the values of the spectral luminance factors of the investigated rock sample can be affected by the angle of incidence i of the light flux, which depends on the angle of the sun, and by the change in the spectral composition of the illuminating flux.

Laboratory observations showed that the nature of the indicatrix of diffusion of the surface of loose sand approximates that of an orthotropically reflecting surface, and hence a change in the angle i should not affect the value of the spectral luminance factors. This view was confirmed by experimental measurements of ρ_λ of different rock samples in field conditions. The photometry of the samples was carried out on the grounds of the Academy of Sciences base in Nebit-Dag (December 18, 1956).

The spectral luminance curves obtained for the investigated specimens are shown in Fig. 19. The curves of ρ_λ for a sample of fine-grained graywacke sand measured at 10 AM, 12 noon, and 2 PM local time were of almost identical shape, and their positions on the coordinate plane (ρ_λ, λ) practically coincided (Fig. 19, sample 18).

Photometry of a specimen of dense calcareous sandstone with the same FM-2 universal photometer in the same conditions and with the same requirements observed as in the photometry of loose sand (sample 18) showed that the spectral luminance curves obtained for this sandstone at 10 AM, 12 noon, and 2 PM local time appeared quite different. The change in the position of the curves on the coordinate plane (ρ_λ, λ) was particularly

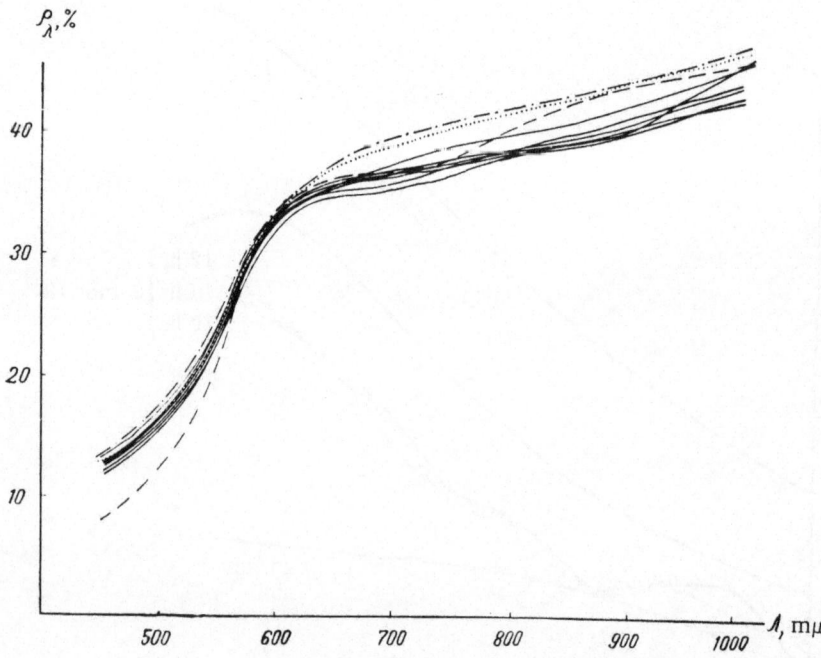

Fig. 20. Spectral luminance curves of sand from a barchan northeast of Ash-khabad. The measurements were made on May 3, 1960 at 1-hr intevals begin-ning from 9 AM local time (dashed line) and ending at 5 PM local time (dot-dash line). The measurements were made with a field spectrometer constructed in 1959 in the Laboratory of Aeromethods and were performed by V.A. Alek-seev and A.A. Boiko.

appreciable. In this case the values of the spectral luminance factors could have been affected by the different type of indicatrix of diffusion of the photometered rock, which exhibits, as was discovered later, a mixed type of reflection with a predominance of the specular component. The values of ρ_λ could also have been affected by the non-neutral nature of the reflection of the investigated sample in relation to the composition of the il-luminating flux.

The effect of the size of the angle of incidence of the light flux on the spectral luminance factors of an investigated object can be eliminated or reduced by choosing the most suitable time of day for photometry. This time can be determined from special tables for calculation of natural light intensity and visibility (Sharo-nov, 1945).

The choice of the time for photometric measurements must be such that the sun's height over the horizon is not less than 45°, in which case the angle i will not exceed 45°; i.e., the nature of the indicatrix of diffusion of the standard (barite paper) and some objects of the sand type will be orthotropic (see Figs. 3-5, 9-11).

In the case of photometry of objects exhibiting strong specular reflection the choice of the time must satisfy other requirements. In this case it is essential to reduce or eliminate highlights, and the photometry process is complicated by the introduction of additional devices, particularly various polaroids — special standardizing devices, insensitive to the change in the values of the angle i, which have still not been suffi-ciently well developed.

The effect of the angle of incidence i of the light flux on the values of the spectral luminance factors of sands was also determined by repeated photometry of the same sample of sand with a field spectrometer. During the photometry the conditions of measurement satisfied all the requirements previously listed. As dis-tinct from the technique of measurements with the FM-2 universal photometer, photometry of the standard and object in this case was not performed simultaneously but in turn, though so rapidly that the conditions of illumination of the standard and object were practically the same.

Photometry of a medium-grained graywacke sand with a field spectrometer on May 3, 1960 in the Ashkhabad region was performed in stationary conditions; i.e., the sample was poured into a flat dish and photometered every hour between 9 AM and 5 PM local time. The instrument was mounted on an open platform and the measurements were made in the sequence: standard – sample – standard. The results of this photometry are illustrated in Fig. 20, which shows that out of the obtained series of almost identical spectral luminance curves the curves obtained at 9 AM (dashed line), 4 PM (dotted line), and 5 PM (dot-dash line) deviated slightly from the rest. In this case the nature of the spectral luminance curves of the photometered sand sample was affected by the change in the angle i, determined by the height of the sun above the horizon (h_0) and, possibly, by the change in the spectral composition of the sunlight.

The next test of the effect of the angle of incidence on the value of ρ_λ was carried out with the same field photoelectric spectrometer, but the objects of the measurements were the natural surfaces of barchan slopes inclined at different angles relative to the incident rays of the sun. The aim of these measurements was to select the most suitable area of a barchan for photometry from the air.

Photometry was first performed on different parts of the barchan; 1) the base; 2) the gentle northwest slope with an angle of inclination of about 20°; 3) the top of the barchan, which has an almost horizontal, even surface; 4) the relatively steep northeast slope, which has an average angle of inclination of 33°. In every case the photometered surface was a clean one, composed of blown sand with traces of wind ripple. The period of time during which the photometry of these four parts was performed was approximately a half-hour between 12 noon and 1 PM local time. The nature of the spectral luminance curves obtained from the photometry of the four listed parts of a barchan with an undisturbed sand surface is shown in Fig. 21 (curves a).

Sand samples taken from the photometered parts of the barchan were then measured again, but this time in stationary conditions; i.e., the sample was poured into a flat dish and the smooth horizontal surface of the sand

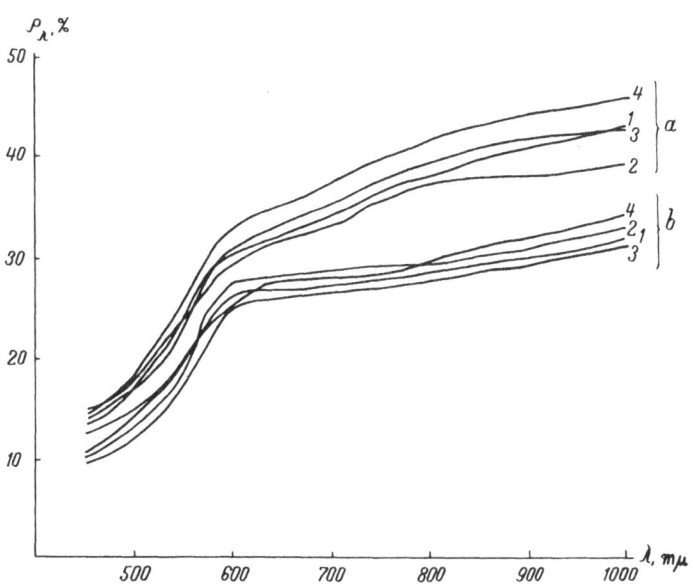

Fig. 21. Spectral luminance curves of sand from different parts of a barchan.
a) Measurements made directly on a barchan oriented with its gentle slope (20°) to the northwest and its steep slope (33°) to the southeast; b) stationary measurements taken from photometered areas of the surface of the barchan: 1) Base of barchan; 2) gentle northwest slope; 3) top of barchan; 4) steep southeast slope of barchan. The measurements were made by V.A. Alekseev and A.A. Boiko on May 6, 1960.

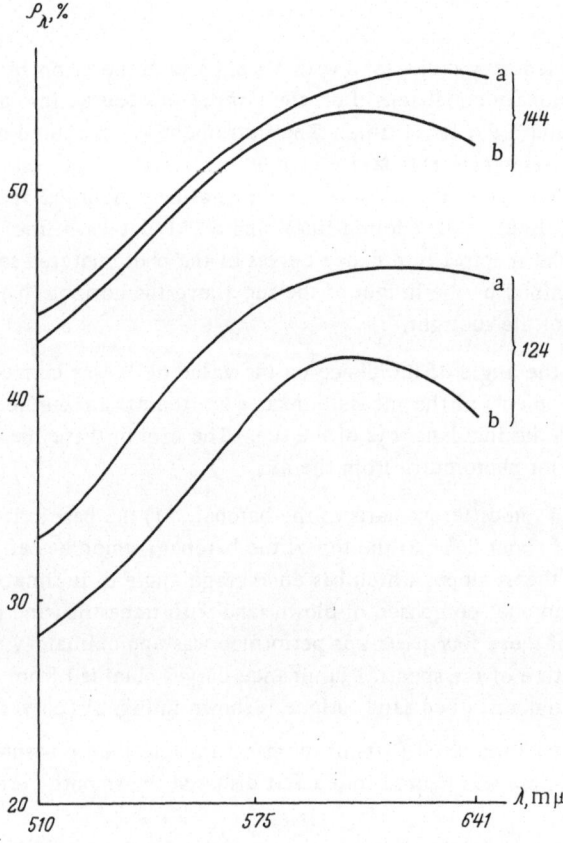

Fig. 22. Spectral luminance curves of sand from a barchan near Burdalyk
(sample 124) and from a barchan south of Chardzhou (sample 144). The measurements were made from the air. The curves a correspond to photometry of
the barchan slopes illuminated by the glancing rays of the sun, and the curves
b are for the sides illuminated almost normally by the incident rays of the sun.

was photometered with the same field photoelectric spectrometer. The conditions of illumination were practically the same as in photometry of the dune (clear, cloudless sky, same time of day). The nature of the curves
of ρ_λ obtained by photometry of sands with a disturbed structure is shown in Fig. 21 (curves b).

A repetition of the measurements of ρ_λ in stationary conditions, where the effect of the changes in the
angle of incidence of the light flux was excluded, showed that in this case the nature of the curves of ρ_λ was
affected by the variation of the granulometric and mineralogical composition of the sands within the barchan
itself due to the eolic differentiation of the fragmentary material comprising the barchan.

The next test of the effect of the angle i on the values of ρ_λ consisted in aerial photometry of barchan
slopes oriented differently relative to the incident sunlight. Figure 22 shows the curves of the spectral luminance
factors obtained by aerial photometry of the steep barchan slopes illuminated by the glancing rays of the sun,
i.e., the "shaded" sides of the barchans (curves a), and the curves obtained by photometry of the gentle slopes
facing the sun (curves b). As the figure shows, the curves are of a different nature. The curves from the steep
slopes, illuminated by the glancing rays of the sun, lie higher in the figure.

The illustrated data show that the relief has an appreciable effect on the value of the spectral luminance
factors. For the avoidance of discrepancies in the measurements of ρ_λ in aerial photometry of sands covering
large areas, the data of which are to be compared, the areas chosen for photometry must be oriented similarly
relative to the incident rays of the sun. In all cases of aerial photometry we measured the reflection spectra
from the tops of gentle barchans, where the surface was almost horizontal. Besides the orientation, another

operative factor here might be the differentiation of the fragmentary material within the barchan. This process is constantly occurring under the influence of winds. Photometry of an area which always occupies the same position on the barchan also assists the obtention of comparable data. It must be pointed out in conclusion that the rate of change of the angle of the sun, as is well known, is different at different times of the day. The rate of change of the angle of the sun is greatest in the morning and evening hours and, hence, it is better not to carry out photometry at these times, otherwise the conditions of illumination of the standard and object will not be the same.

Effect of mist and dust on spectral luminance of object. The mist due to the presence of water vapor in the air has practically no effect on the values of ρ_λ of investigated objects in the case of photometry from heights of less than 100 m. Moreover, in work in sandy desert regions, where the water vapor content of the air varies around 8-10%, the effect of such mist is practically negligible.

In the case of photometry from low heights, the scattering of reflected radiant energy will mainly be due to the presence of a dust haze. A particularly dense dust haze is produced when a helicopter is used for the work. In this case the haze consists mainly of relatively large particles (about 1.0 mµ), which according to Minnaert (1954) scatter rays of all wavelengths equally. The formation of dust hazes is usually accompanied by an increase in wind and general unfavorable meteorological conditions. In such a case aerial spectrophotometry of the investigated objects must be abandoned.

Effect of clouds in sky at time of photometry on values of ρ_λ. The presence of various kinds of clouds in the sky at the time of photometry has a pronounced effect on the illumination of the investigated surface, on the spectral composition of the incident radiant flux and, hence, on the values of ρ_λ.

The effect of clouds on the values of ρ_λ of an investigated object was tested by repeated photometry of the same sand sample in different cases: under a completely clear sky (Fig. 23a), when the sky was partially covered with thin cirrus clouds (Fig. 23b), and when the sky was covered with high layered clouds (Fig. 23c). A comparison of the spectral luminance curves obtained for the same sand sample shows that the curves obtained by photometry when the sky was covered with continuous layered clouds lay a little higher than the curves obtained by photometry under a completely clear sky. The shapes of the compared curves also differed slightly from one another (Fig. 23).

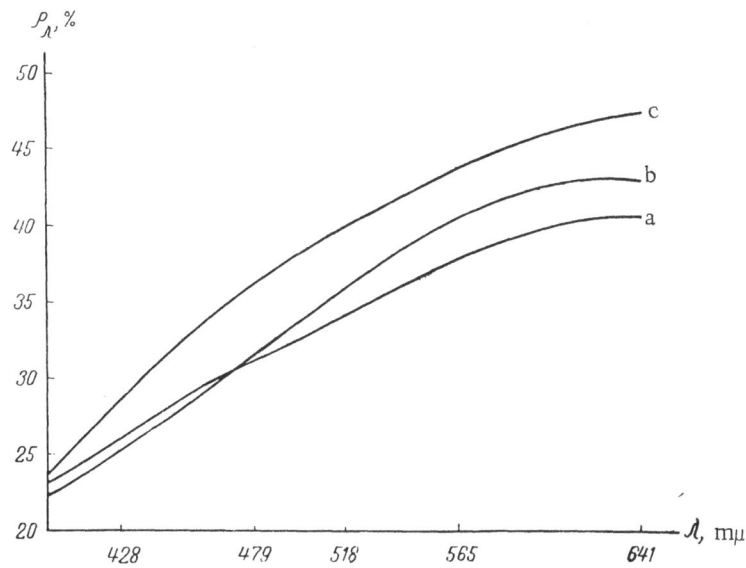

Fig. 23. Spectral luminance curves obtained for a sample of calcareous sandstone by photometry with the FM-2 universal photometer in the following conditions: a) Clear cloudless sky; b) sky partially covered with light cirrus clouds; c) almost all the sky covered with high thin-layered clouds.

For elimination of the errors due to the presence of particular clouds at the time of photometry, the measurement of ρ_λ of the investigated objects must be carried out only under a clear, cloudless sky.

Effect of Nature of Photometered Surface on Values of ρ_λ

By the nature of the surface we mean the degree of roughness or microrelief of the photometered surface and the degree of dispersion of the particles comprising the measured area.

The roughness or smoothness of a photometered surface was evaluated by Sytinskaya, who introduced an additional factor q, called the smoothness factor, into photometric practice. This parameter characterizes the law of scattering of light by the given surface and is greater the greater the optical smoothness of the surface. Sytinskaya gave a table of values of smoothness factors. Some values from this table characterize objects considered in our investigations:

Values of Smoothness Factor
(according to N.N. Sytinskaya)

Barite paper	+0.91
Milk glass	÷1.30
Quartz sand	+0.14
Loamy soil	+0.20
Rough limestone	+0.07
Crystalline salt crust	+2.74
Amorphous salt crust	+0.81
Moss	-0.07

Thus, q is unity for orthotropically reflecting surfaces, q > 1 for glossy surfaces, q < 1 for rough surfaces, and q is less than zero for highly divided surfaces (Sytinskaya, 1946).

In the case of photometry from a height, the surface of rocks will almost always have some degree of roughness and this roughness will be registered on the spectrograms. For instance, the spectrogram obtained by photometry of an inhomogeneous surface (various objects) has distinct longitudinal bands (Fig. 24). Microrelief in the form of wind-ripple marks, where the distance between the crests of the ripple is approximately 10 cm and the height of the crest is about 1 cm, is registered on the photograph of the object and affects the nature of the spectrograms, which have fine longitudinal strips (Fig. 25) in the case of photometry from a height of 10-15 m.

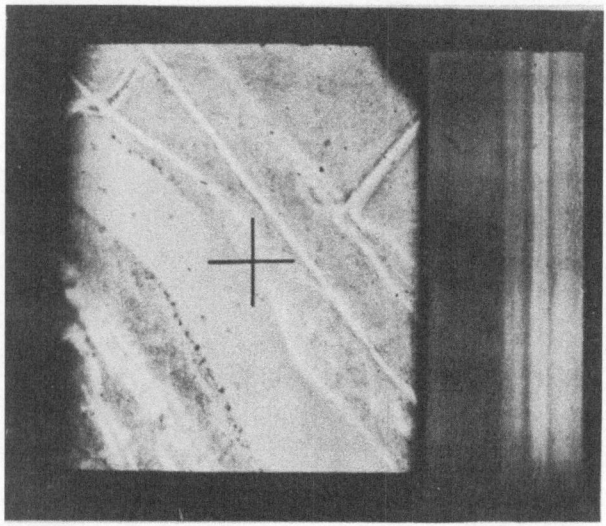

Fig. 24. Longitudinal stripes on spectrogram obtained by photometry of inhomogeneous objects (water and sand).

To determine the nature of the effect of the orientation of wind-ripple marks on the value of the spectral

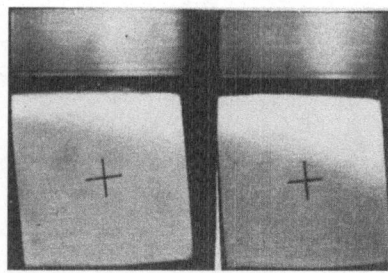

Fig. 25. Thin longitudinal stripes on spectrogram obtained by photometry of a barchan surface covered by wind-ripple marks.

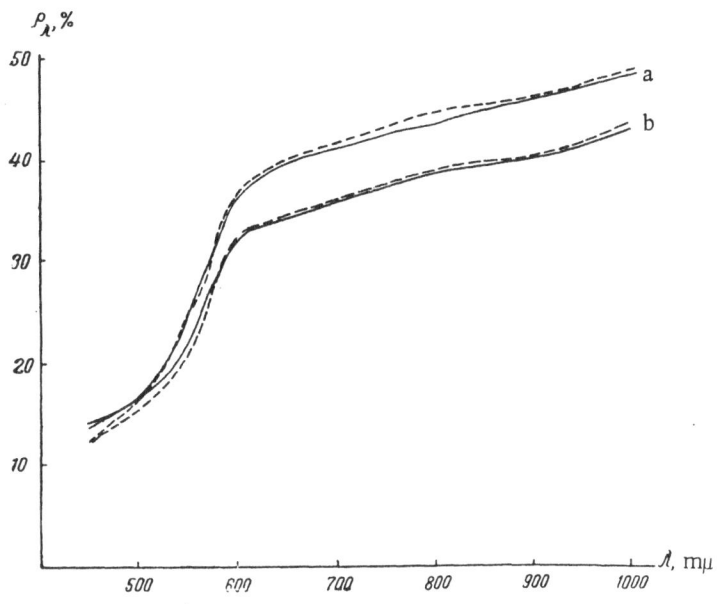

Fig. 26. Two cases (a and b) of photometry of the surface of sand barchans with different orientation of wind-ripple marks. Solid lines show results for one orientation, and broken lines show results when orientation of wind-ripple marks was turned through approximately 90° relative to the first case.

luminance factors, we carried out special measurements on the same part of a barchan, but with different orientation of the ripple marks relative to the plane of symmetry of the photometer. Figure 26 illustrates two cases of measurements where the wind-ripple marks were oriented parallel to the optical plane of the spectrograph (curves shown by broken line) and perpendicular to it (curves shown by continuous line). The measurements made in this way confirmed the view that in the case of photometry of orthotropically reflecting surfaces the orientation of the microrelief has no effect on the values of the spectral luminance factors.

The nature of the photometered surface depends on the degree of dispersion of the material comprising the measured surface. The effect of the degree of dispersion of a material on the values of its spectral luminance factors has been studied by various investigators. Zhidkova conducted a series of experiments in which she measured the diffuse reflection factors of colored glass pulverized to give different degrees of dispersion. As a result of her measurements (Zhidkova, 1954) she found that weakly absorbing substances in the dispersed state have higher absolute diffusion reflection factors the greater their dispersion (Fig. 27).

Data on the photometry of single-mineral fractions of various degrees of dispersion are given in the works of Tolchel'nikov and Belonogova (1959). Figure 28 shows some curves taken from the works of these investigators. As in Zhidkova's works on the measurement of the diffuse reflection factors of powdered colored glass, the authors concluded that "the spectral luminance of minerals increases with their fragmentation." This conclusion is definitely correct, but if we consider the curves of the spectral luminance factors of three quite different minerals (see Fig. 28) [transparent colorless quartz (pulverized single crystal of rock-crystal), pink microcline, and dark mica (biotite)], then we see that each mineral has a series of curves parallel to one another, corresponding to each of the photometered fractions, and that for each different mineral the more highly dispersed fractions have higher spectral luminance factors. In this case the smoothness factor of the photometered surface, which affects the height of the position of the curve of ρ_λ in the coordinate system (ρ_λ, λ), is evidently the important factor. Hence, from these examples we can draw the very important conclusion that the spectral course of the spectral luminance curves of each of the investigated minerals remains the same.

The objects of our investigations were sands, which contain a mixture of fragments of varying size and different mineralogical composition. If such a mixture of fragmentary material is passed through sieves, the obtained fractions will differ in mineralogical composition. The fine fractions will usually be rich in accessory

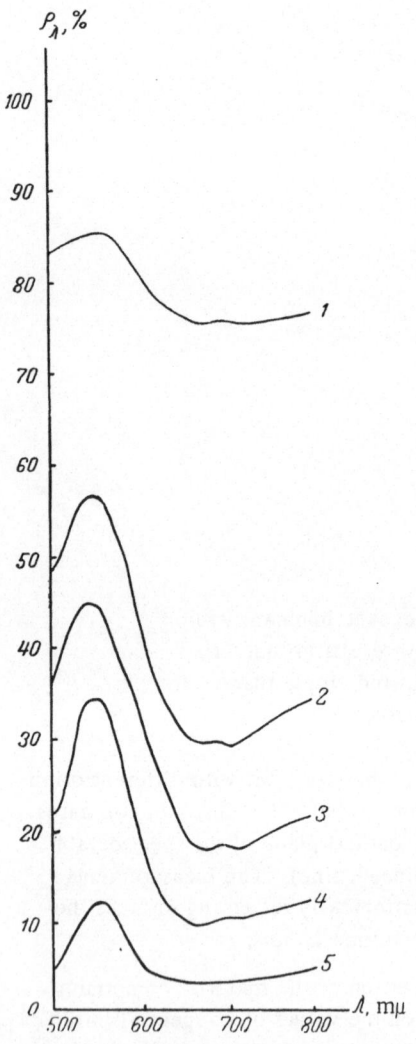

minerals, which have higher refractive indices and, accordingly, different indices of reflection. To test this hypothesis we photometered coarse (0.21-0.29 mm) and fine (0.05-0.07 mm) fractions of sands of different petrographic composition: a typical graywacke sand from the region of Burdalyk (Fig. 29) and a predominantly quartz sand from the Tedzhen delta (Fig. 30). In the first case the position of the curve of the fine fraction, which was rich in zircon, was above that of the coarse fraction, which was rich in rock fragments. In the second case, where the coarse fraction consisted of quartz grains with a considerable admixture of kyanite, the spectral luminance curve was approximately at the same level as that of the fine fraction (see Fig. 30).

Thus, in the photometry of deposits of the type of modern sands it is necessary to take the nature of the surface into account and to try to measure horizontal areas, such as the tops of gentle barchans. This partially eliminates errors due to the effect of the smoothness factor. Moreover, the effect of the differentiation of the fragmentary material according to its degree of dispersion is minimal on flat areas.

The investigation of the effect of the properties and mineralogical composition of rock on the values of its spectral luminance factors is the subject in this book. The problem in this case reduces to the obtention of comparable measurements. The values of ρ_λ obtained on measurement of the investigated rocks must be independent of the conditions of photometry and the properties of the standard with which the reflecting properties of the investigated object are compared.

Hence, the observation of the following strict rules of photometry is of very great importance:

1. The conditions of illumination in the photometry of object and standard and of compared objects must be identical. To ensure this, photometry in the field should be carried out:

    a) in clear cloudless weather in the absence of any dust haze;

    b) at fixed hours of the day and on a surface close to

Fig. 27. Spectral luminance curves of fractions of pulverized ZhZS-1 glass (according to Zhidkova). 1) 0.001 mm; 2) 0.044-0.053 mm; 3) 0.105-0.125 mm; 4) 0.210-0.250 mm; 5) 1.00-1.190 mm.

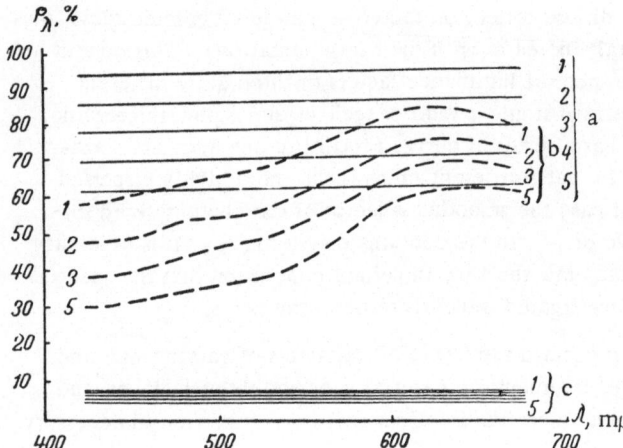

Fig. 28. Spectral luminance curves of single-mineral fractions pulverized and passed through sieves to give the following fractions: 1) < 0.1 mm; 2) 0.10-0.25 mm; 3) 0.25-0.50 mm; 4) 0.50-1.0 mm; 5) 1.00-3.00 mm. a) Quartz – rock-crystal; b) microcline; c) biotite. (Tolchel'-nikov's data, 1960).

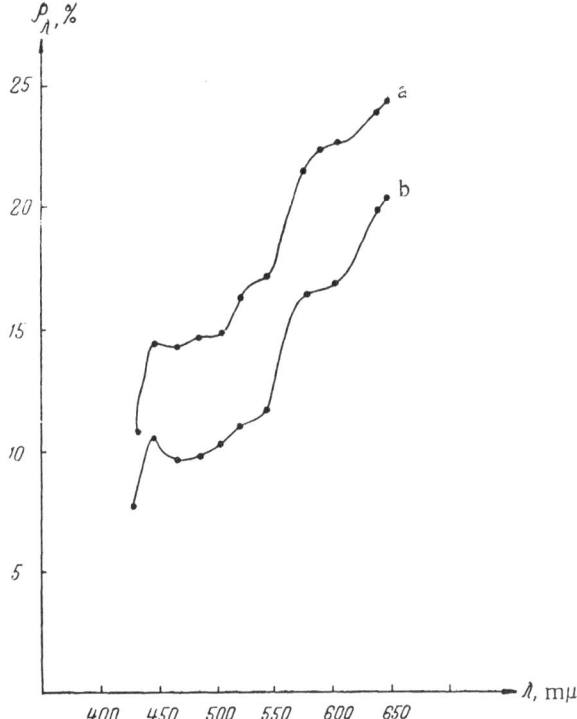

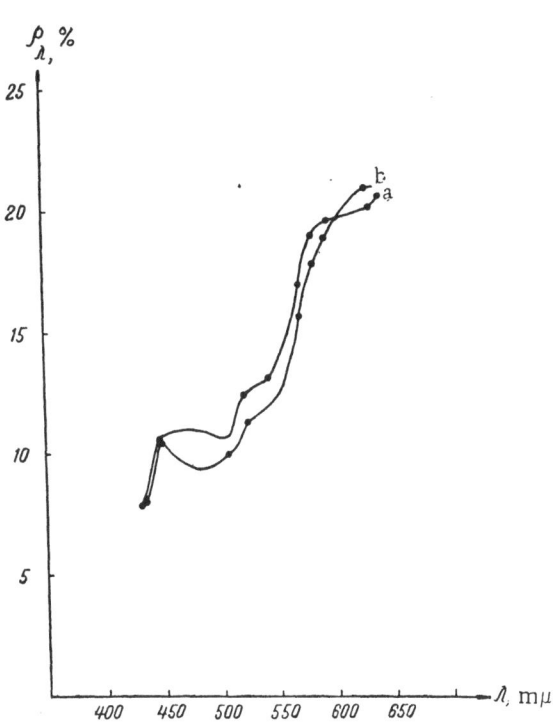

Fig. 29. Spectral luminance curves obtained for graywacke sand from the region of Burdalyk (sample 124) by photometry of fractions: a) 0.053-0.07 mm; b) 0.21-0.29 mm.

Fig. 30. Spectral luminance curves obtained for predominantly quartz sand from north of Tedzhen delta (sample 18) by photometry of fractions: a) 0.05-0.07 mm; b) 0.21-0.29 mm.

horizontal so that the angle of incidence i of the radiant flux does not exceed 45° (in view of the strong effect of the angle i on the indicatrix of diffusion of the object and standard):

c) in the nadir position, where the angle ε between the registered reflected beam and the perpendicular to the measured surface is equal or almost equal to zero.

2. The nature of the photometered surfaces should as far as possible be the same. The areas for photometry should be smooth, free from any other formations, and the surface of the outcrop must be fresh, with no traces of weathering, desert varnish, salt efflorescences, etc., or the nature and extent of weathering processes must be indicated.

When the above conditions of photometry are observed, the obtained spectral luminance factors or other parameters characterizing the type of the curves of ρ_λ can be compared with the data of petrographic and mineralogical analyses of the investigated rocks. The discovered relationship can be used to estimate the occurrence of particular types of rocks on the photometered area. The compilation of maps of ρ_λ values of rock, converted to petrographic composition, is aeropetrographic mapping.

CHAPTER III

METHODS OF MEASURING SPECTRAL
LUMINANCE FACTORS OF ROCKS

This chapter gives a brief account of the principal instruments used for measurements of the reflecting properties of rocks. The purpose of describing the instruments is merely to acquaint the geologist with the operating principle of each instrument and the conditions in which it is used. Such knowledge is essential for efficient utilization of the apparatus.

Instruments for measuring the reflecting properties of rocks can be divided into two large groups: laboratory and field instruments. We will describe them below in accordance with this classification.

Laboratory Instruments for Measuring the Reflecting Properties of Natural Objects

Laboratory instruments for measurement of the reflecting properties of natural objects include indicatometers and various types of photometers. Representatives of this group are the indicatometer of Leningrad State University and the FM-2 universal photometer, which are discussed below.

Laboratory instruments for measuring the nature of scattering of radiant energy also include pyranometers, which are based on the registration of the thermal energy obtained by conversion of the reflected radiant energy. These instruments have been known for a long time and have been applied for practical measurements of various natural objects by different investigators (Coblenz, 1912; Ångström, 1919; Kalitin, 1929, 1931). The use of instruments of this type for rock photometry will require special improvements in design, probably involving the incorporation of semiconductor thermoelements. This promises to provide highly sensitive recording instruments, which we do not yet have at our disposal.

Below we give descriptions of typical laboratory instruments which we have used for measuring the spectral luminance factors of rocks and, in particular, of the investigated sandy deposits.

The Indicatometer of the Astronomical Observatory of Leningrad State University

Indicatometers are instruments for measuring the indicatrices of diffusion of light energy reflected from the investigated surface of an object.

The basic idea in the construction of indicatometers is to provide a means of measuring the luminance factors of an investigated object at different angles of incidence i of the light flux and at different angles ε determining the position of the photometer registering the reflection from the investigated surface. In the case of measurement of oriented surfaces, the azimuths of the incident rays a_i and the azimuths of the reflected rays a_ε must be known.

The indicatometer of Leningrad State University has been described in the literature (Radlova, 1943; Orlova, 1958). This indicatometer is a typical representative of such instruments and, hence, we will give a brief description of it.

The indicatometer consists of an object stage on which the specimen is placed an illuminating device, and a photometer mounted at the end of a rotating arc. In this case, as in all other indicatometers, there are three mutually perpendicular axes (Fig. 31):

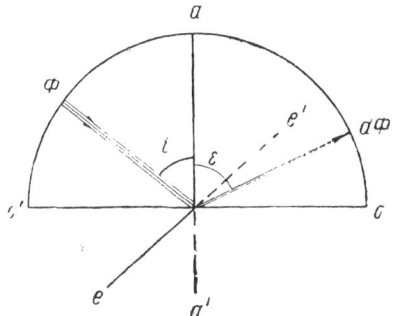

Fig. 31. Design principle of indicatometer. aa') Axis of rotation of stage with specimen; this axis is perpendicular to the surface of the specimen; oo') axis of rotation of lamp; ee') axis of rotation of photometer registering reflected energy; Φ) light flux incident at angle i; dΦ) light flux reflected at angle ε.

1) The axis of rotation aa' of the object stage. This axis is perpendicular to the surface of the specimen (the specimen must be mounted with the investigated surface exactly horizontal so that the axis aa' is perpendicular to the investigated surface).

2) The axis of rotation oo' of the lamp, which is mounted on the end of a large arc; by rotation of the arc the angle of incidence of the light flux on the investigated surface can be accurately set; the angle i can be varied from 0 to ± 90°.

3) The axis ee' is the axis of rotation of the photometer, which analyzes the intensity of the light beam reflected from the investigated surface.

In the indicatometer of the Astronomical Observatory of Leningrad State University the axis is stabilized. The axis of the object stage, on which the specimen for measurement is mounted, and the vertical axis of rotation of the lamp coincide with the axis of the instrument. The point of intersection of the axes aa' and oo' is the main center of the instrument. The light flux reflected from the investigated surface falls on a totally reflecting prism and is then measured by a visual surface photometer.

The construction of this indicatometer provides for rotation of the beam with the photometer in only one plane. The arc with the lamp has two degrees of freedom, i.e., it can rotate around the axis oo' and around a perpendicular to this axis. The standard is a screen, which is measured under standard conditions of illumination with angles i = 0°, ε = 50°, and the difference of azimuths $a_i - a_\varepsilon = 0°$. In the visual field of the photometer lens the surface luminance of the investigated specimen is seen against the background of the luminance of the visual field, and by rotation of the photometer drum the observer equalizes the luminances of these fields in each individual case for all positions of the lamp and photometer.

According to requirements the indicatrices of diffusion can be obtained: 1) for a fixed angle of incidence of the light flux (angle i constant) and varying angles of the measured reflected flux (angle ε varying from +90 to -90°); 2) for a fixed angle ε and angle i varying from +90 to -90°.

The technique of measurement of luminance indicatrices on this indicatometer after its adjustment has been checked consists of the following operations:

1) The specimen for measurement is placed on the object stage, which has leveling screws.
2) The zero point of the indicatometer lamp is determined by taking readings on the circle.
3) The beam with the photometer is set in one of the extreme positions, where the angle ε = +10 or −10°.

The values of the luminance factors of the investigated surface are taken every 5 or every 10° according to requirements. It is more convenient to perform the measurements with the beam carrying the photometer in a fixed position and varying angles of the lamp: i = 10, 20, 30, 40, 50, 60, 70, 80, and 90°. The beam should then be turned around its axis through 5 or 10°, and another series of measurements of the luminance of the investigated surface at different angles i is carried out.

4) The luminance factors for the construction of the indicatrices of diffusion are calculated from the formula

$$\varrho = \varrho_s \frac{B_i}{B_0} \sec i,$$

where ρ is the luminance factor of the investigated object; ρ_s is the luminance factor of the standard; B_i is the luminance of the object for an angle of incidence equal to i; B_0 is the luminance of the screen when i = 0.

According to the data of Orlova (1958), the luminance of the investigated object measured on this indicatometer is calculated from the formula

$$\log B = \log B_0 \pm q\,(x - x_0),$$

where q is a constant equal to 0.032 in measurements on this indicatometer; x is the reading obtained in measurement of the investigated object; x_0 is the reading for the standard.

When the values of the antilogarithms of the luminance factor have been obtained, polar diagrams are plotted on some selected scale. In these diagrams lengths proportional to the values of the luminance B_i are marked off on each radius vector of the corresponding angle i. The obtained polar diagram is the indicatrix of the luminance factor of the investigated object for the particular position of the measuring photometer.

The results of measurements of the indicatrices of the luminance factors of a specimen of granophyric granite from Krasnovodsk Peninsula are given as an example (Table 2). The obtained indicatrices of these factors for a fixed position of the photometer are illustrated in Fig. 32.

Indicatrices of luminance factors can be obtained by measurement of the investigated objects on devices of the type used by Ostrovskii (1956) (Fig. 33). In such devices the reflected radiant flux is registered by a photocell mounted on a special rotating support.

Any spectrometer equipped with appropriate axes of rotation can be used as an indicatometer. For instance, the field spectrometer described below, equipped with a special axis of rotation and associated reading scale, can be used to obtain the indicatrices of spectral luminance factors of an investigated object.

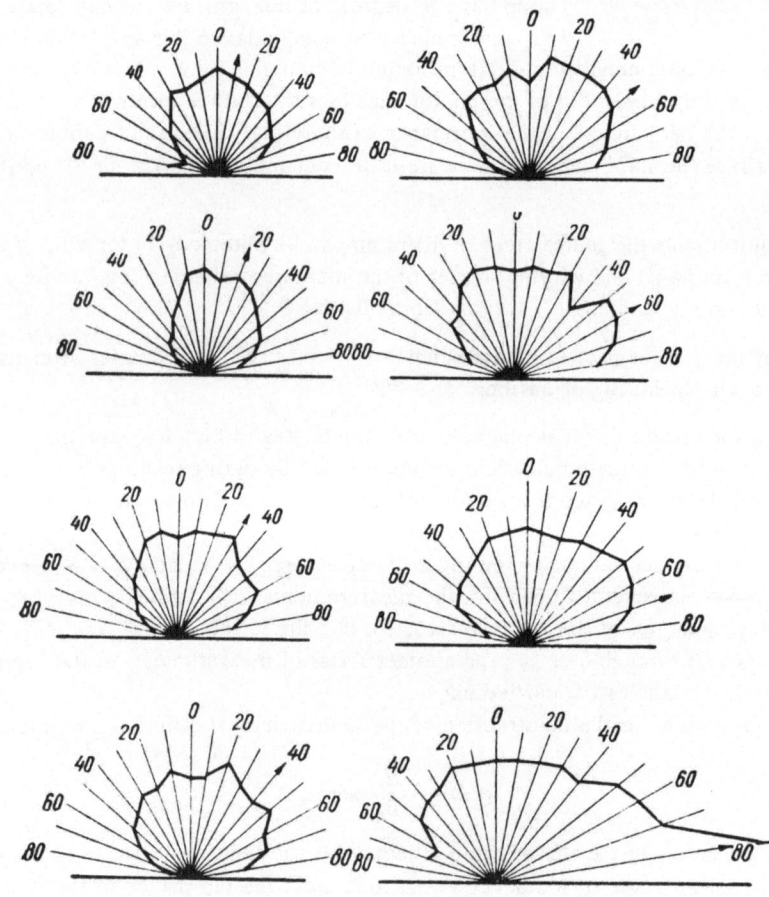

Fig. 32. Indicatrices of luminance factors of a specimen of granophyric granite from Krasnovodsk Peninsula (Ufra).

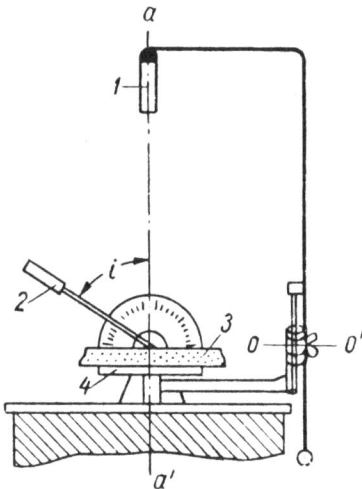

Fig. 33. Diagram of device for determination of indicatrices of luminance factors (according to Ostrovskii, 1956). 1) Photocell; 2) lamp; 3) specimen; 4) rotating stage; oo') axis of rotation of photocell; aa') axis of rotation of stage; i) angle of incidence of light flux.

The FM-2 Universal Photometer

The FM-2 universal photometer has been in use for several years for the photometry of geological objects, principally for measurements of the spectral luminance factors of individual rock samples in stationary or laboratory conditions (Danchev, 1947, 1956; Vistelius and Yaroslavskaya, 1954; Romanova, 1956, 1958).

A general view of the FM-2 universal photometer is shown in Fig. 34. The instrument is based on the same general principle as all other photometers; the equalization of two light fluxes reflected from two different surfaces, one of which has been adopted as a standard.

The optical system of the photometer is shown in Fig. 35. The two light fluxes from the standard I and the specimen II pass through the objectives 5 and the rhombic prisms 6 to the biprism 7, which brings the two light fluxes onto the axis of the eyepiece 8. In the field of the eyepiece the observer sees a circle divided into two halves. The brightness of the left half of the field depends on the light flux reflected from the standard placed on the right and, correspondingly, the brightness of the left half of the circle depends on the light flux reflected from the object placed on the left. Diaphragms 1, 2 with variable apertures are connected with the measuring drums 3, 4. A light filter can be inserted between the biprism 7 and the eyepiece I', II'. The whole set of filters is carried in a special disk (D, Fig. 34). The required filter is selected by rotation of this disk. The brightnesses of the fields are equalized by rotation of the measuring drum (Dr, Fig. 34). The divisions on the drum give the percentage ratios of the area S of the diaphragm aperture at the corresponding opening to the area S_0 of the fully open diaphragm aperture. Rotation of the drum decreases the aperture through which the light flux reflected from the standard passes, thus reducing the brightness of the standard field and making it equal to the brightness of the

TABLE 2*. Data of Measurements of Indicatrices of Luminance Factor of a Specimen of Granophyric Granite from Krasnovodsk Peninsula*

Angles of incidence(i) of light flux, deg	Angle of photometer (ε) registering reflected flux, deg							
	10	20	30	40	50	60	70	80
80	25.0	6.5	6.0	8.0	7.0	8.0	11.0	
70	12.5	11.0	15.0	16.5	20.0	20.5		31.5
60	20.0	15.0	20.0	25.0	28.5		36.0	31.0
50	30.0	22.0	26.5	37.0		40.0	40.0	46.5
40	36.0	24.0	33.5		48.0	41.5	43.0	50.0
30	41.5	32.0		46.5	48.0	52.0	46.0	52.0
20	41.5		50.0	46.5	48.0	56.0	52.0	58.0
10		50.0	50.0	52.0	54.0	56.0	56.0	58.0
0	52.0	52.0	48.0	50.0	56.0	54.0	58.0	60.0
10	48.0	48.0	52.0	50.0	60.0	56.0	56.0	60.0
20	46.5	50.0	54.0	60.0	58.0	56.0	56.0	62.5
30	45.0	48.0	56.0	52.0	58.0	60.0	60.0	67.0
40	41.5	41.5	48.0	50.0	58.0	65.0	62.5	67.0
50	36.0	36.0	46.5	52.0	54.0	62.5	67.5	78.0
60	28.5	32.0	46.5	46.5	46.5	60.0	65.0	84.0
70	22.0	24.0	38.5	41.5	41.5	54.0	60.0	90.0
80	10.0	13.5	28.5	33.5	34.0	41.5	48.0	135.0

*The table gives the antilogarithms of the luminance factors divided by two. The measurements were performed on the large indicatometer of Leningrad State University.

field corresponding to the investigated object. The values of the measured luminance factors ρ_λ are read off from the dial of the drum. They are given as percentages relative to the standard:

$$\varrho_\lambda = \frac{S}{S_0} \cdot 100\% .$$

The brightnesses of the fields corresponding to the standard and investigated object are equalized at specified wavelengths, which are secured by the insertion of a particular filter. The properties of the filters and their optical charactersitcs are known beforehand, having been chosen with due regard to the reflection characteristics of the investigated objects. For the visible region of the spectrum, filters with transmission regions spaced as far as possible at equal wavelength intervals are usually used. The construction of a light filter for an accurately prescribed wavelength presents some difficulty and, hence, the wavelength intervals between adjacent filters are not always equal. Moreover, none of the filters transmits a single precise wavelength, but a small range of wavelengths − the transmission width. In the construction of the filters it is essential to secure the smalles possible width of transmission band. The filters usually employed with the universal photometer are made of glass coated with a thin film of gelatin containing an organic dye and with another glass layer cemented on top for protection against damage. Gelatin filters have low stability and wide transmission bands and are not suitable for accurate measurements.

In all our laboratory measurements we used interference filters, which operate on the principle of reflection of radiation in the whole region except the required wavelength. Interference filters consist of several layers of transparent dielectrics* chosen so that the light interference gives the maximum transmission in the required spectral region. Each interference filter is characterized by the spectral curve of its transmission factor.

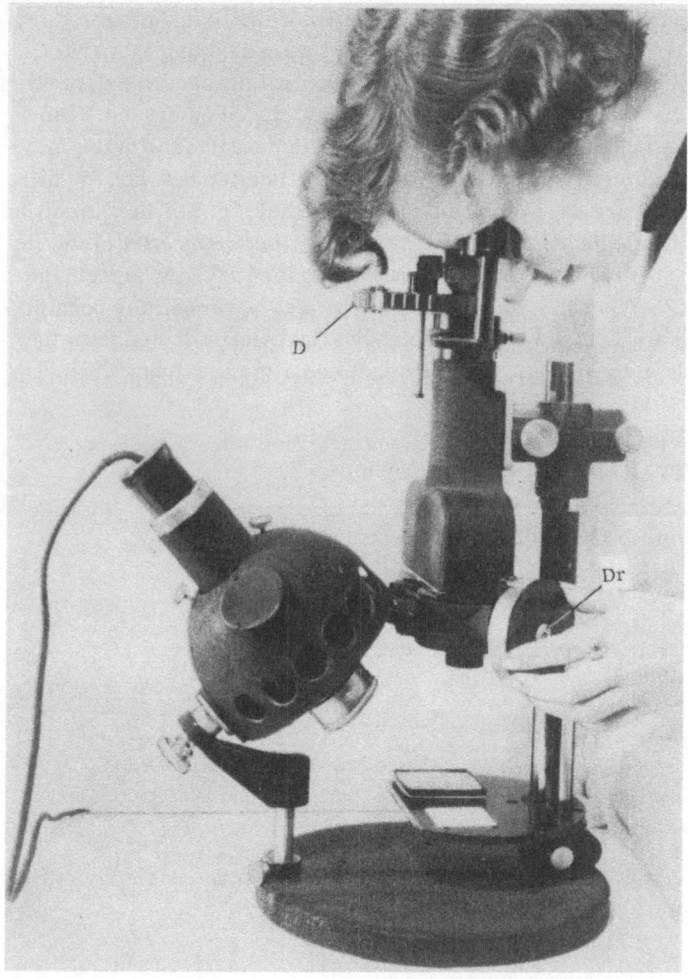

Fig. 34. General view of FM-2 universal photometer. Dr) Drum for equalizing brightnesses of fields of reflection from object (gray) and standard (white); D) disk with set of interference filters. Photo by A.D. Mikhailov.

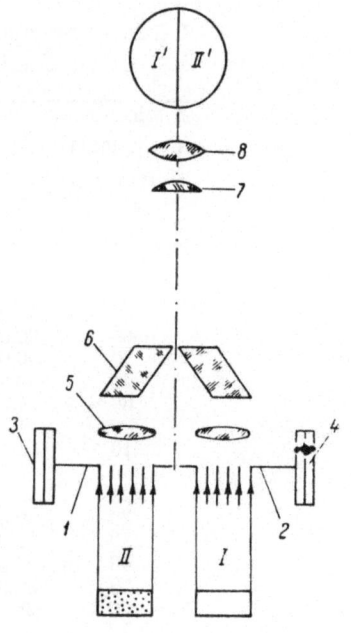

Fig. 35. Optical system of the FM-2 universal photometer.

* Materials used for the several layers of transparent dielectrics are titanium dioxide with n = 2.2 and silicon dioxide with n = 1.45.

36

Table 3. Characteristics of Set of Interference Filters Used in Conjuction with the FM-2 Universal Photometer.

Wavelength, λ_{Tmax}, mμ	Maximum transmissivity T, %	Transmission width $\Delta\lambda$, mμ
398	33.5	12
428	25.0	14
447	36.0	12
467	33.5	12
479	33.0	13
507	24.0	12
518	24.0	12
539	28.0	13
565	29.0	12
577	20.5	11
593	27.0	12
629	28.0	14
641	32.5	12
680	22.0	12

As an example, Table 3 characterizes a typical set of filters constructed in the State Optical Institute in 1956. The service life of interference filters is relatively short and they require periodic testing.

With an adequate set of filters the universal photometer can be used to measure the values of the spectral luminance factors of a rock sample over the whole visible region of the spectrum. The values of the spectral luminance factors can be expressed either as a percentage of some standard or in absolute values ρ_λ. In the latter case the values of the spectral luminance factors obtained by measurement (ρ_λ, %) must be multiplied by the luminance factor of the standard used for comparison. In practice it is sufficient to obtain the relative values (ρ_λ, %), from which the spectral luminance curve of the measured object is plotted. Conversion to absolute values only affects the position of the curve in the system of coordinates (ρ_λ, λ, mμ).

In measurements with the FM-2 universal photometer, the sample and standard are in absolutely identical conditions of illumination. In laboratory conditions the light source is an incandescent lamp emitting a beam of light at an angle of 45° to the horizontal surfaces of the investigated object and standard. The sample is photometered several times. Practice has shown that three measurements of each ρ_λ are sufficient and the sample should be rotated through approximately 45° in the plane of the stage after each mesurement. The mean value of these three measurements is used to plot the spectral luminance curve of the sample. An example of the measurement of some rock samples and the plotting of the curves of the spectral luminance is given in Table 1 and in Fig. 16.

The spectral luminance factors of rocks can be measured with the universal photometer in the field by the photometry of separate small samples in sunlight. For this purpose the sample is mounted directly on the stage of the instrument, as shown in Fig. 34. By using objectives with different focal lengths and fitting a support of appropriate height, it is possible to photometer individual parts of rock outcrops.

The photometered area is a square of side m. The size of the photometered area depends on the height H of the objective above the investigated object and is calculated from the formula

$$m = 2H\tan\beta,$$

where β is the viewing angle of the instrument (the FM-2 has β = 12°). In photometry in the laboratory, where the sample is placed on the stage of the photometer, m = 2 cm; i.e., the photometered area is 4 cm^2.

Measurement of one rock sample on the universal photometer takes an experienced observer about 7 min. Hence, it is quite understandable that photometry by means of this instrument is impossible from any moving vehicle, to say nothing of photometry from an airplane or helicopter.

In view of this it became necessary to design special instruments – spectrometers, which enable rapid photometry of small areas of the earth's surface from the air.

Field Spectrometers*

The main requirements of field spectrometers are rapidity of recording reflection spectra, stability of operation, and durability in transport. The purpose of these instruments is the photometry of small parts of the surface of natural objects on the ground or from the air.

---

* All field spectrometers can be used for measurements in laboratory conditions.

Among field spectrometers, various spectrographs (with glass and quartz optics) for recording reflection spectra on any photographic layer have been employed in practical work for a long time.

Spectrometers based on the principle of photoelectric registration of reflection spectra have recently been employed in field spectrometry. In these instruments a combination of photocell and photomultiplier provides a means of measuring small radiant fluxes. There are various methods of recording the measured radiant fluxes: directly on a photographic emulsion (spectrovisors), or by a pointer-type ammeter with an appropriate reading scale (field photoelectric spectrometers).

A description of some field spectrometers which have been tested in field conditions is given below.

An example of a spectrometer based on the principle of photographic registration of the reflection spectrum is the RShch-1 aerial cinespectrograph.

The RShch-1 Aerial Cinespectrograph

In aerial photometry the instrument for registering the radiant energy reflected from a small area of the earth's surface must register it rapidly during the flight and must provide a tie-up between the obtained spectrum and the investigated outcrop. These problems have been solved by the use of aerial spectrographs. The first types of aerial spectrograph had a large angle of view, thus enabling the obtention of the integral luminance of the area, and registered the image very slowly. The spectrograms obtained with these spectrographs could not be interpreted.

The unsatisfactoriness of these instruments for geological research induced us to undertake the design of a special instrument which would provide a stable spectrum of the reflection from the investigated rock, record this spectrum reliably, and simultaneously check that the spectrum was obtained from a clean surface by providing a photograph of the external appearance of the photometered area. We constructed and tested an instrument of this type in 1957. The optical part of the instrument was designed in accordance with our technical specifications by Yu. P. Shchepetkin of the State Optical Institute.

The RShch-1 aerial cinespectrograph is designed for the photometry of small areas of rocks from the air. A diagram illustrating the registration of the spectrogram and the photograph of the photometered area is shown in Fig. 36. The photometered area σ with sides M and K occupies the center of the photographed area, which is a square of side L. A special spectral system with objective f_c is used for photometry, and a separate, parallel optical system with objective f_i is used for photography. Both systems are focused on the same frame l (area 16 × 32 mm) of the film. A general view of

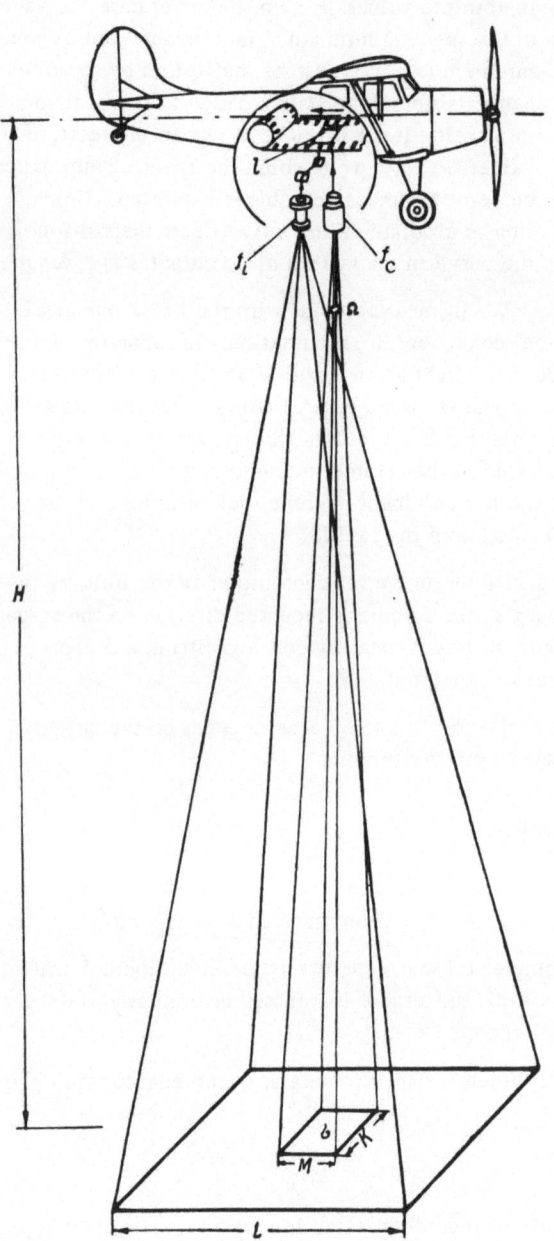

Fig. 36. Diagram illustrating aerial photometry with the aerial cinespectrograph.

Fig. 37. General view of RShch-1 aerial spectrograph.
C) Cinecamera; S) spectrograph objective; L) objective
of landscape system (Romanova, 1960).

the cinespectrograph is shown in Fig. 37, where the cinecamera C is detached from the spectrograph, which is attached to the optical system for photography of the locality. The optical system of the instrument is shown in Fig. 38.

The photographic system 1-5 consists of an objective 1 (Industar-22, 1 : 3.5, f = 51.4 mm), a collector 2 (f = 40.4 mm) consisting of two plano-convex lenses between which are enclosed the diaphragm of the landscape field and cross-wires for marking the center of the spectrophotometered part on the locality. The turning system [lens 3 (f' = 51.0 mm) and two mirrors 4 and 5] directs the image plane given by the objective 1 onto the emulsion surface of the film in the film aperture 12. The landscape image occupies the upper part of the frame and has an area of 15 × 15 mm.

The spectral system 6-11 consists of a condenser 6, a slit 7, which gives an image of the photometered area, a plane mirror 8 with light dimensions 19 × 26 mm, and a collimator 9. The dispersing element here is a flat reflecting diffraction grating 10 of the echelette type with 600 lines/mm. This grating concentrates the bulk of the light energy in the plus first-order spectrum at an angle of incidence $\varphi_0 = -31°36'$. The angles of diffraction for K = +1 are:

$$\lambda_1 = 434.1 \text{ m}\mu \quad (\text{ line } g'), \quad \varphi_1 = 51°\ 40';$$
$$\lambda_0 = 546.1 \text{ m}\mu \quad (\text{ line } e), \quad \varphi_0 = 58°\ 24';$$
$$\lambda_2 = 656.3 \text{ m}\mu \quad (\text{ line } c), \quad \varphi_2 = 66°\ 36'.$$

The light dimensions of the grating are 45 × 50 mm.

The objective 11 of the Jupiter-9 spectrograph camera (aperture ratio 1:2, f' = 85 mm, 2β = 28°) is inclined so that the spectrum is focused on the film in the best way. The image of the spectrum occupies the lower part of the fram and is 6 × 15 mm. The viewing angle of the RShch-1 aerial cinespectrograph is 3°, and the depth of focus is 20 m to infinity. The photometered region of the spectrum covers the range from 496 to 662 mμ. The spectral characteristic of the instrument, obtained by measurement of the photographed spectrum of iron, is illustrated in Fig. 39, which shows that the characteristic is almost linear.

A general view of a frame of film with the photograph of the locality and the spectrum of the photometered area is shown in Fig. 40. The area marked on the photograph of the locality by a cross indicates the region from which the reflection spectrum is obtained. The shape of the photometered area depends on the width of the spectrograph slit, but in general form it is a rectangle oriented with its short side M perpendicular to the direction of flight.

The size of the photographed area (with side L) depends on the height of flight H, on the focal

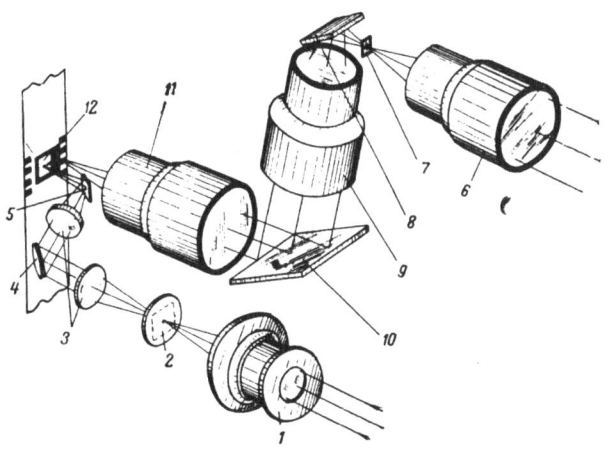

Fig. 38. Optical system of RShch-1 aerial cinespectrograph.

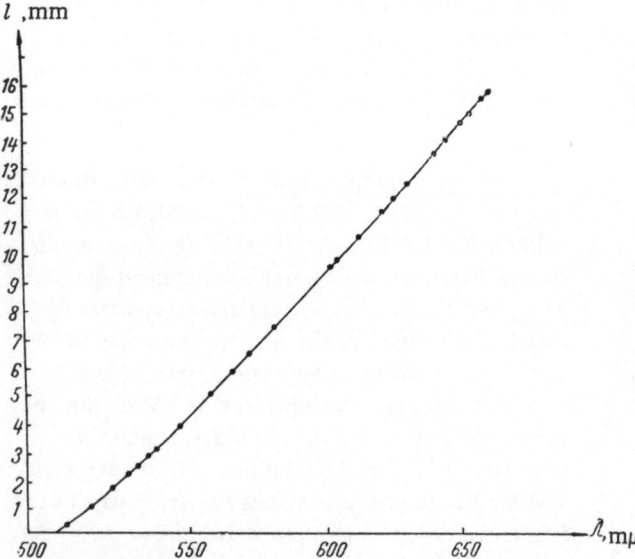

Fig. 39. Linear spectral characteristics of aerial cine-spectrograph RShch-1 from measurement of photographed spectrum of iron.

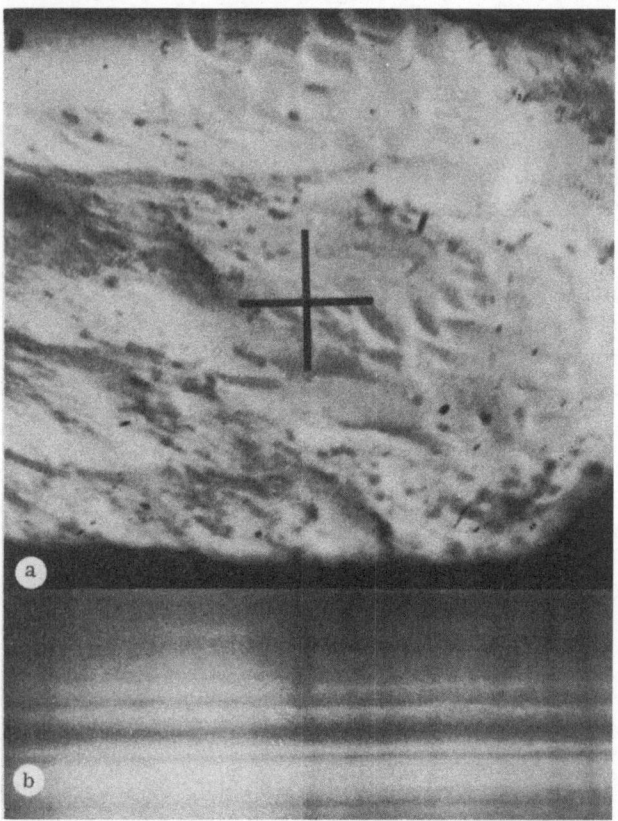

Fig. 40. Photograph of (a) surface of photometered region (landscape) and (b) reflection spectrum from the central area marked by a cross on the photograph.

length f_i of the landscape objective and on the dimension l of the negative on the frame of the film. This relationship is given by the formula

$$L = l \frac{H}{f_i} .$$

In addition, the blurring Δ' of the picture which occurs in photography from a moving airplane must be taken into account. This blurring is more pronounced the greater the flight speed (v, m/sec) and the lower the camera speed (n, frames per sec). The displacement Δ of the axis of the objective during the exposure of one frame is equal to

$$\Delta = \frac{v}{2 n} .$$

The blurring is calculated from the formula

$$\Delta' = \Delta \frac{f_i}{H} ,$$

and, hence, substituting the value of Δ, we obtain

$$\Delta' = \frac{v}{2 n} \cdot \frac{f_i}{H} .$$

In calculation of the size of the photometered area, the width δ of the spectrograph slit must be taken into account as well as the flight altitude H and the focal length f_c of the spectrograph objective. The slit width is set by a special screw before the start of the work.

The above-indicated (see Fig. 36) width M and length K of the photometered area are determined from the following formulas:

$$M = \delta \frac{H}{f_c} , \quad K_1 = \Delta'' + \delta \frac{H}{f_c} ,$$

where K_1 is the total length of the area spectrometered during one exposure; Δ'' is the shift in the plane of the spectrograph slit, equal to the displacement f_c of the condenser axis multiplied by the scale: $\Delta'' = \Delta \cdot f_c / H$.

An example of a calculation of the dimensions of the photographed and photometered areas in work with the described cinespectrograph mounted on an airplane with an average flight speed of 120 km/hr for the case of a camera speed of 16 frames/sec and a spectrograph slit width equal to 0.193 mm is given in Table 4.

The total weight of the cinespectrograph is about 6 kg. It is portable, convenient for transportation, and rugged in operation. This has been

40

TABLE 4. Dimensions of Surveyed Area in Relation to Flight Altitude

Flight altitude, m	Photographed area		Photometered area		
	area L^2, m^2	blurring Δ', mm	length K, m	width M, m	area σ, m^2
10	8.51	5.35	0.71	1.06	0.75
20	34.40	2.67	1.41	1.09	1.54
100	851.47	0.54	7.06	1.28	9.04
1000	85147.24	0.05	70.59	3.39	239.30

confirmed in three years of work in the arduous field conditions of sandy deserts with numerous landings on the barchans.

Photometry of the surface of rocks consists in comparing the reflection spectrum of the investigated object with the reflection spectrum obtained from a standard under the same conditions of illumination. Owing to the design of the instrument, photometry of the standard can only be carried out on the ground. For the obtention of spectrograms of comparable photographic density, the standard is photometered either with neutral filters with an accurately measured transmission density or with an inserted stepped optical wedge.

The standard, photometered with an optical wedge in front of the spectrograph slit, gives a band spectrogram (Fig. 41). Each band of the spectrogram is regularly reduced by a known attenuation factor. The presence of a series of bands on the spectrogram from the standard enables a choice of the one closest in density to the spectrogram of the investigated object.

The stability of operation of the aerial cinespectrograph ensures the obtention of reliable values of spectral luminance factors. As shown earlier (see Fig. 39), the spectral characteristic of the instrument is almost a straight line and, hence, an accurate analysis of the observations can be made. The mean square error of the measurements with allowance for the errors occurring in the processing of the exposed film and in measurement of the photographic densities on the microphotometer does not exceed 5%.

Fig. 41. Spectrogram of standard photographed with an inserted optical wedge.

Fig. 42. This iron emission spectrum, obtained in work with an aerial cinespectrograph, is used to determine the limits of the spectrum or the working region of the spectrograph.

The technique of field work with the aerial cinespectrograph consists in the following:

1. The cinespectrograph is mounted on the airplane or helicopter so that it can be easily re-removed for photometry of the standard as soon as the aircraft lands.

2. Photometry of the standard is carried out directly before takeoff and after every 1.5-2 hr of surveying. The standard (barite paper) can be kept in a special cassette, which is opened at the time of photometry. The standard is photometered with an optical wedge inserted and with a fixed slit width. The camera speed in the photometry of the standard and objects must be the same. It is convenient to have a special stand for holding the cinespectrograph during photometry of the standard.

3. After the standard has been photometered the spectrograph is mounted again in the hatch of the airplane (helicopter) and the aerial photometry of the investigated objects begins.

In the run over the object, the cleanness of the rock surface to be photometered and the direction of the airplane's shadow, which may cover the photometered area, must be noted.

During photometry of the investigated objects it is also necessary to ensure that the correct adjustment of the spectral and landscape systems is maintained. (The adjustment of these systems is carried out in the laboratory and then the two systems are rigidly fastened together.) The adjustment can be checked by total or partial covering of the photometered area with the shadow of the aircraft. If the shadow covers a region close to the area marked by the cross on the photograph of the locality, the spectrogram is reduced in width, and when the shadow of the aircraft covers the whole photometered area the spectrogram disappears altogether.

The wavelength interval registered by the spectrogram is set during the general adjustment of the spectrograph. For this purpose the emission spectrum of iron was photographed (Fig. 42). From the spectral lines of iron known from the atlas (Zaidel' et al., 1952) the boundaries of the corresponding wavelengths of the spectrogram are determined.

The faults inherent in all aerial spectrographs which register reflection spectra photographically and in the described aerial cinespectrograph are as follows:

1. The times of photometry of the standard and the investigated object do not coincide. The interval between these times introduces errors due to changes in the conditions of illumination of the compared objects (standard, measured rock), and leads to some distortion of the values of ρ_λ.

2. The accuracy of photographic photometry is about 5-10% on the average (Prokof'ev, 1951). The processing of the obtained spectrograms is laborious, and the rate of processing of these spectrograms is low.

3. Photographic photometry is used to register reflected light energy, i.e., to investigate the spectral luminance of objects only in the visible region of the spectrum. Yet the nature of the reflection of radiant energy, particularly in the infrared region, may give additional characteristics of the reflecting properties of rocks.

For removal of the above faults, for acceleration of the photometry of standard and object, and for more accurate determination of the spectral luminance factors of natural objects, aerial spectrometers based on the principle of photoelectric registration of spectra have been designed. Descriptions of one type of aerial spectrovisor and one ground-type field photoelectric spectrometer are given below.

Spectrometers Employing Photoelectric Registration of the Spectrum

In the aerial photometry of any rock outcrop it is very important that the photometered area be homogeneous in composition and that its surface be clean and free from any additional formations. Homogeneity of composition and cleanness of the rock in the investigated area can be much more easily secured if a small area is photometered.

A reduction in the size of the photometered area when aerial spectrographs are used can be achieved by one or more of the following measures:

a) the use of long-focus objectives in front of the entrance slit of the spectrograph;
b) a reduction in the width of the spectrograph slit;
c) an increase in the speed of registration of the spectrum.

All these measures lead to an appreciable reduction in the amount of radiant energy incident on the focal plane of the spectrograph. The energy of the spectrum is insufficient to cause decomposition of the silver halides in the photosensitive layer of the film, i.e., photographic registration of the spectrum is unsuitable in this case. This quite naturally leads to a search for other, more sensitive methods of registering reflection spectra.

Vacuum photoemissive cells have been found to be suitable radiation receivers in conditions of low light intensity. The photocurrent of these devices is easily amplified electronically. A photocell with subsequent multistage secondary-emission amplification (photomultiplier) provides a means of measuring small radiant fluxes, down to 10^{-9} lm or even less (Volosov, 1960).

Numerous descriptions of spectrometers based on the principle of photoelectric registration of the spectrum have recently appeared in the Soviet literature (Berkovich et al., 1959; Boiko, 1959; Voronkova et al., 1960;

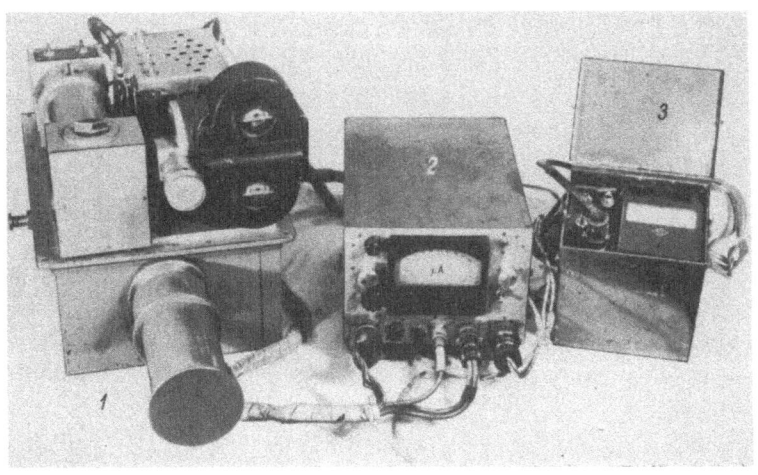

Fig. 43. General view of spectrovisor outfit. 1) Monochromator; 2) supply unit; 3) batteries for supplying incandescent lamp of illuminators. Photograph by K. E. Meleshko.

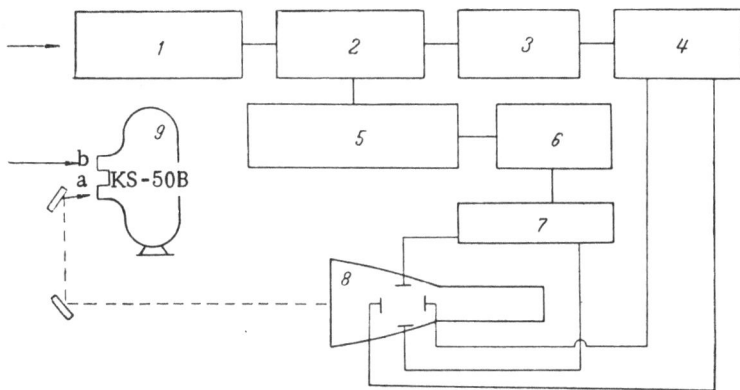

Fig. 44. Block diagram of 1959 model spectrovisor. 1) Monochromator for separation of the radiant flux by wavelength; 2) scanning device, consisting of an oscillating mirror mounted on the armature of an electromagnet, for scanning the spectrum in time; 3) radiant-energy receivers consisting of two FEU-17 and FEU-22 photomultipliers connected in parallel and covering the wavelength region 400 to 1000 mμ; 4) vertical amplifier for deflection of the beam of the cathode-ray tube along the y axis, i.e., producing a deflection proportional to the intensity of the reflected radiant flux; 5) synchronizing device for triggering the sweep generator at the instant corresponding to a particular position of the oscillating mirror; 6) sweep generator for sweeping the beam horizontally across the screen; 7) horizontal amplifier deflecting the beam of the cathode-ray tube along the x axis, i.e., over the wavelength range; 8) cathode-ray tube with screen displaying a curve characterizing the spectral distribution of the reflected radiant energy; 9) cinecamera for photographing the curve from the screen of the cathode-ray tube (lens a) and the image of the photometered surface (lens b).

Kol'tsov,1959, 1960; Krasil'shchikov et al., 1958; Mironova and D'yachkova, 1957). Only a few types of these spectrometers have been designed for aerial spectrophotometry and can be used for the photometry of geological objects from airplane (helicopter).

In 1956, Kol'tsov (1959) developed a model of an aerial spectrovisor with a glass prism. Two FEU-17 and FEU-22 photomultipliers connected in parallel were used as receivers and provided for measurement in the wavelength range 400 to 1000 mμ. The wavelength distribution of the radiant energy was displayed in the form of a curve on the screen of a cathode-ray tube, and this curve was then photographed with a cinecamera. Wavelength scanning of the spectrum was effected by a linear reversible movement of the entrance slit over the spectrum and gave fairly reliable results in photography at a speed of 20-25 frames/sec. An increase in the scanning rate resulted in some instability of the curves (oscillograms) owing to the strong vibrations. Kol'tsov subsequently improved his model and introduced a new and much lighter version (Kol'tsov, 1960) with a diffraction grating operating in the same region of the spectrum (400-1000 mμ).

A model of a spectrovisor of approximately the same design of those of Kol'tsov was fitted with additional new devices and was briefly described in the papers of Lyalikov et al. (1958) and Voronkova et al. (1960). This spectrovisor has been used in our photometric investigation of geological objects in the field from a helicopter. This spectrovisor is described below as an example of a spectrometer employing photoelectric registration of the spectrum.

The 1959 Model Spectrovisor Constructed by a Group of Workers in the Laboratory of Aeromethods (Voronkova et al., (1960)

A spectrovisor is a fast electronic spectrometer designed for aerial spectrophotometry of small areas of the earth's surface. A general view of the complete outfit of the described spectrovisor is shown in Fig. 43. The main components of this model of spectrovisor are shown in the form of a block diagram in Fig. 44.

Figure 45 illustrates the spectrovisor optical system, which operates in the following way. The radiant flux reflected from the investigated surface arrives at the objective 1 of the spectrometer, passes through the entrance slit 2 and the collimator lens 3, and falls as a parallel beam on the diffraction grating 4, which has 600 lines/mm and an area of 20 × 20 mm. The spectrally decomposed radiant flux falls on the externally aluminized oscillating mirror 5 and after reflection from this mirror passes through the lens 6, which magnifies the image of the exit slit. Lenses 9 and 10 image the entrance slit on the photocathodes of the photomultipliers 12 and 13: an FEU-17 photomultiplier with an antimony—cesium photocathode for the region 350-450 mμ and an FEU-22 with a cesium oxide—silver photocathode for the red and near-infrared region of the spectrum (450-1100 mμ). The electrical signal (proportional to the intensity of the monochromatic radiant flux) is amplified and applied to the vertical deflection plates of the cathode-ray tube. System 14-20 provides for the formation of two marks indicating the boundaries of the spectrum on the screen of the cathode-ray tube. System 21-26 is for calibration of the instrument by photography of the spectrum of a light filter (a PS-7 glass), which has characteristic absorption bands. The light signal from lamp 21 passes through lens 22, is stopped down by a diaphragm, then passes through lens 23, is directed onto the filter 25, and falls on the exit slit. The Dove prism 26 can open and close the exit slit to the flux coming from the objective 1. With the prism in the upper position, the slit transmits the beam of light from objective 1, and with the prism in the lower position the slit transmits the beam passing through the filter 25 from the incandescent lamp 21. The synchronous shutter 27 is for marking the zero line on the screen of the cathode-ray tube.

The employment of FEU-17 and FEU-22 electron multipliers connected in parallel as a registering system provides for operation of the spectrovisor in a wide region of the spectrum. Figure 46 shows the spectral characteristics of these photomultipliers. This graph shows that these curves overlap one another in the 420-600 mμ region. An OS-11 light filter (28 in Fig. 45) is introduced to exclude the superposition of the second-order spectrum in the 800-1000 mμ region. The photomultipliers are supplied by a stabilized voltage of 1000-1200 V ± 0.1% when the input voltage varies by −15 to +10%.

For the supply of the oscillograph and audio oscillator the direct voltage of the aircraft system, equal to 27 V, must be converted to an alternating voltage by means of a motor alternator.

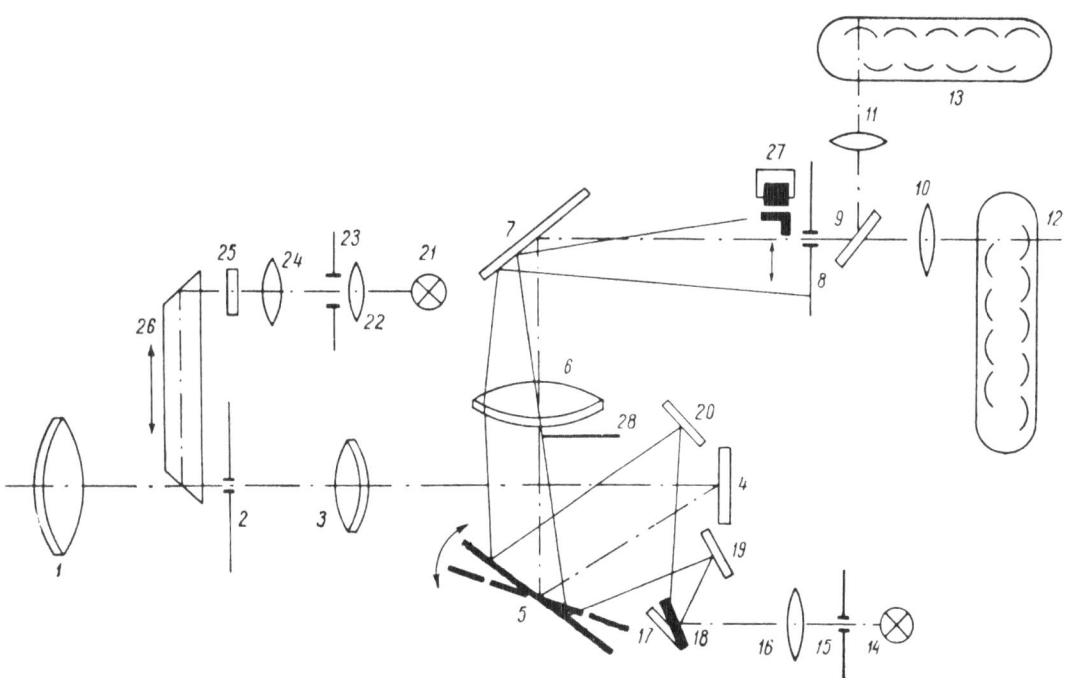

Fig. 45. Optical systems of 1959 model spectrovisor.

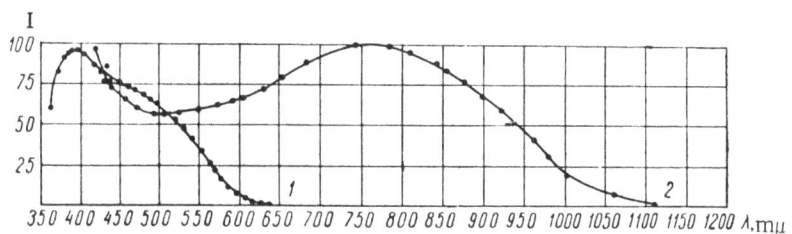

Fig. 46. Spectral characteristics of photomultipliers: 1) FEU-17; 2) FEU-22.

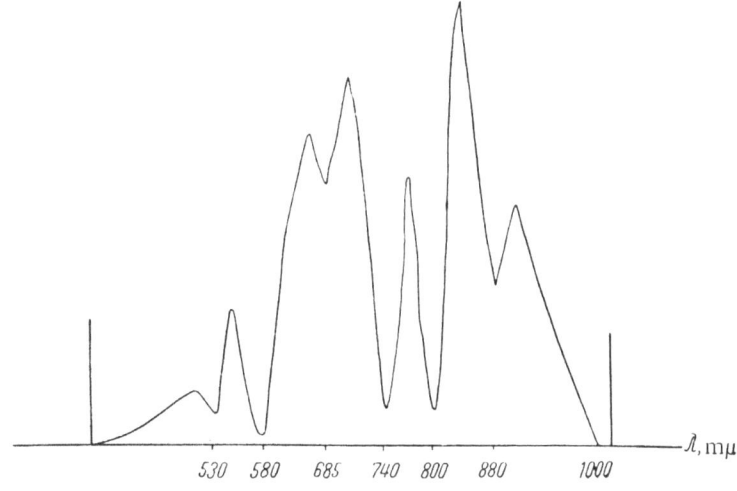

Fig. 47. Position of the PS-7 filter transmission minima, which are used for wavelength calibration of the spectrovisor curve.

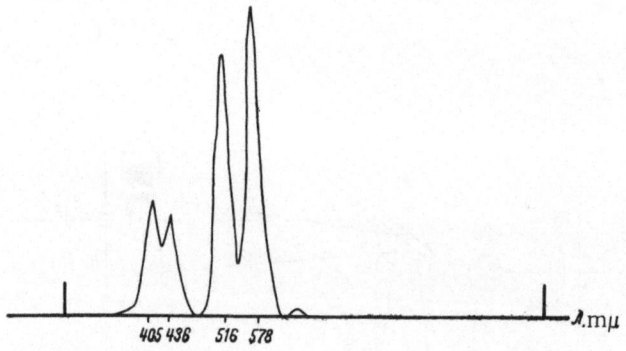

Fig. 48. Spectral distribution of radiant energy in the case of photography of the spectrum of a UFO-4A mercury lamp.

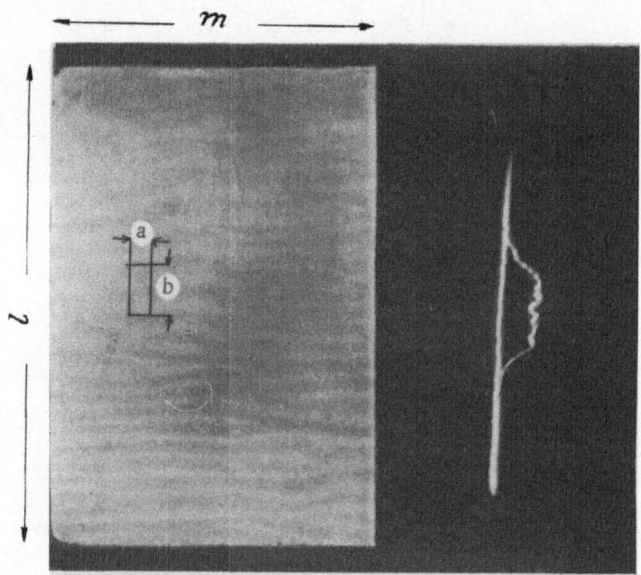

Fig. 49. General view of cinefilm frame on which the photometered area is photographed, and the curve (oscillogram) characterizing the distribution of radiant energy reflected from this area.

TABLE 5. Dimensions of Photometered Area in Relation to Flight Altitude

Flight altitude H, m	Length of sides of photometered area	
	length a, m	width b, m $\left(S\dfrac{H}{f} + vt\right)$
10	0.11	0.015 +0.42
100	1.13	0.15 +0.42
150	1.70	0.23 +0.42
300	3.40	0.46 + 0.42
400	4.52	0.60 +0.42

Thus, the reflecting properties of the investigated object, measured by means of the 1959 model aerial spectrovisor, are recorded in the form of curves characterizing the spectral distribution of the reflected energy according to the properties of the object. This curve is traced on the screen of the cathode-ray tube by the electron beam, which is acted on by two voltages: a) the vertical voltage, which is taken from the load resistance of the photomultipliers and is proportional to the intensity of the monochromatic radiant flux; b) the voltage taken from the sweep generator, which is synchronized with the oscillatory movement of the mirror; this voltage causes the horizontal deflection of the beam, i.e., the wavelength scanning of the spectral curve.

The wavelength stability of the calibration curve of the spectrovisor is checked by photography of a PS-7 filter and the spectrum of a mercury lamp in the laboratory and in flight. The calibration curve is plotted from the position of the transmission minima of the PS-7 filter (Fig. 47) and from the mercury lines (Fig. 48).

The curves of spectral distribution of the radiant energy are recorded by filming the screen of the cathode-ray tube with a KS-50 cinecamera. The same camera is used to photograph the locality containing the photometered area through a specially mounted additional lens 9b (see Fig. 44). A general view of a frame obtained in photometry with the 1959 model spectrovisor is shown in Fig. 49, where l and m are the sides of the photographed area, and a and b are the sides of the photometered area.

This model of spectrovisor, like all other single-beam spectrometers, can only measure relative spectral luminances. This means that during the work successive measurements of a standard and the investigated object must be carried out. Instead of barite paper a special standardizing device (Fig. 50) is used to reduce to a minimum the time between photometry of the standard and photometry of the object. This standardizing device is rigidly attached to the frame for mounting the spectrovisor in the door opening of the airplane (or helicopter). The spectrovisor is fitted so that it can rotate around a horizontal axis and the spectrometer objective can be directed downwards onto the photometered object and then upwards into the standardizing device (in Fig. 50 the spectrometer objective is pointed upwards and inserted in the tube of the standardizing device).

In aerial photometry the spectrovisor is mounted so that the entrance slit of the monochromator is perpendicular to the line of flight. In this case the spectrophotometered area will be oriented with its long side a parallel to the entrance slit, and its short axis b perpendicular to the entrance slit. In Fig. 49, side a appears shorter owing to the greater displacement of the condenser axis during the exposure of one frame (vt).

The dimensions of the photometered area are given by the following formulas:

$$a = h\,\frac{H}{f}, \quad b = S\,\frac{H}{f} + vt,$$

where H is the flight altitude; v is the flight velocity; f is the focal length (265 mm) of the monochromator lens; t is the time for the obtention of one spectral curve (approximately 1/120 sec in this model of spectrovisor); h is the height of the entrance slit of the spectrometer; S is the width of the spectrometer slit.

Table 5 gives the dimensions of the spectrophotometered area in relation to the flight altitude for the case where the flight velocity is 180 km/hr, the height of the spectrometer slit is 3 mm, and the slit width is 0.4 mm.

The area occupied by the landscape photograph on the frame is 16 × 11.5 mm; the area occupied by the oscillogram on the frame is 16 × 10 mm.

The size of the photographed area when a spectrovisor with a Jupiter 11 objective (f = 135 mm, relative aperture 1:4) is used is very small. For instance, in photography from a height of 10 m, in view of the size of the negative (16 × 11.5 mm) of the photographed area, the dimensions of the photographed area will be

$$l = 0.016 \cdot \frac{10}{0.135} = 1.18\,\text{m};$$

$$m = 0.0115 \cdot \frac{10}{0.135} = 0.85\,\text{m};$$

i.e., in work from a height of 10 m only the cleanness of the photometered surface can be assessed from the photography of the locality. The photograph cannot provide a reference to the topography of the locality.

TABLE 6. Means $\overline{X}$ and Standard Deviations S of Values of Spectral Luminance Factors of Sands from Region of Lowland Kara-kum Obtained by Measurement in the Field with the 1959-Model Spectrovisor (data of 1960)

Wavelength, mμ	Number of measured outcrops (sand samples)							
	480		495		498		499	
	$\overline{X}$	S	$\overline{X}$	S	$\overline{X}$	S	$\overline{X}$	S
460	0.4	1.3			7.8	3.6		
480	1.3	2.7	9.2	3.4	13.3	4.2	8.2	1.7
500	4.4	2.8	12.9	4.3	16.9	2.2	10.8	1.9
520	8.5	3.0	16.7	4.3	20.5	3.3	15.1	2.5
540	11.4	3.2	19.5	4.0	23.4	3.3	19.4	2.7
560	14.8	3.1	22.5	3.8	26.6	3.2	21.5	3.2
580	18.3	2.8	26.1	4.6	30.7	3.3	25.3	3.7
600	20.5	2.9	29.1	5.1	33.6	2.9	28.7	4.1
620	21.9	2.7	30.5	4.6	35.6	2.9	29.7	3.9
640	22.1	3.2	31.1	5.3	36.0	2.9	31.1	6.0
660	22.5	3.3	32.3	4.9	36.2	3.1	32.5	6.5
680	22.0	3.9	32.1	4.7	35.1	2.7	29.9	5.8
700	22.0	3.8	30.2	2.4	31.3	1.6	28.6	5.8
720	22.2	4.2	31.5	4.9	30.0	1.5	27.8	5.6
740	22.8	4.9	33.1	4.2	30.4	1.9	28.0	7.3
760	23.4	5.5	34.1	5.6	31.2	2.3	28.1	7.1
780	24.0	5.7	33.7	6.4	31.8	3.3	28.5	5.2
800	21.1	4.3	31.9	7.4	28.3	3.2	26.9	5.2
820	21.4	5.2	32.8	7.1	30.8	3.5	28.5	7.1
840	22.5	5.4	32.9	6.7	33.3	4.2	29.8	7.4
860	21.2	4.8	31.7	6.7	33.6	5.3	29.0	9.5
880	18.8	4.4	29.2	6.2	30.6	8.5	26.7	7.4
900	14.8	8.0	24.0	8.8	28.0	11.4	26.1	8.9
Number of observations	19		15		9		10	

There is practically no blurring in photography of the landscape. This is achieved by a reduction of the shutter opening in front of this part of the frame to secure an exposure of 1/300 sec.

General description of 1959-model spectrovisor. 1. The spectrovisor can be used to measure the spectral luminance of small areas of investigated objects in the wavelength range 450-950 mμ.

2. The resolution of the instrument, determined from the half-width of a mercury spectrum line, is 20 mμ.

3. The mean square error of measurement of ρ_λ, including the errors in measurement of the oscillograms, is 2-3% in laboratory conditions. In field spectrophotometry from a helicopter this error is much larger, particularly in the infrared part of the spectrum (Table 6).

4. The time for recording of a spectrum is 1/120 sec.

5. The current consumed by the spectrovisor from the supply system of the airplane (or helicopter) is 10 A.

6. The wavelength calibration curve of the instrument preserves its linear character and the displacement relative to the marks indicating the boundaries of the spectrum on the oscillograph screen does not exceed 15 mμ.

7. The weight of the whole outfit is 110 kg. If the audio oscillator is replaced by a semiconductor oscillator, the weight of the apparatus can be reduced to 75 kg.

When this model of spectrovisor is compared with other instruments for aerial photometry of small areas of any objects, the following points should be noted:

1. The spectrovisor can spectrophotometer areas of much smaller size than the aerial cinespectrograph.

2. The existing model of spectrovisor permits spectrophotometry in the wavelength range 450-950 mμ, i.e., it can be used for aerial investigation of the reflecting properties of rocks in the near-infrared region.

3. The existing model of spectrovisor provides for frequent measurement of the standard. This greatly improves the accuracy of the obtained values of ρ_λ.

Fig. 50. Standardizing device D fitted on the objective of the spectrophotometer S for photometry of the standard in the air.

The following features must be regarded as drawbacks of this model:

1. The great weight of the whole structure.

2. The complexity of the design and the presence of moving parts, especially the plane mirror of the scanning device, which make the instrument liable to damage and to frequent breakages during its operation, especially in field conditions.

3. The instability of the operation of semiconductor components when the instrument is used in different climatic conditions, such as desert conditions, where the temperature in the cabin of the helicopter varies around +50°, means that the semiconductor oscillator must be replaced by an audio oscillator.

4. The curves of the spectral distribution of radiant energy are traced by a thick beam; this can lead to additional errors in the processing of the oscillograms.

5. The conversion from oscillograms to spectral luminance factors is laborious and time is wasted in measuring the ordinates with a stencil.

6. The asymmetrical position of the whole assembly on board the airplane (helicopter) relative to the plane of symmetry of the aircraft introduces some navigational difficulties in the run over the area to be spectrophotometered.

For a check of aerial measurements of spectral luminance factors by the described model of spectrovisor, and allowance for the effect of vibrations, dust haze, and other factors, the spectrophotometry of the investigated deposits can be duplicated by ground measurements of the spectral luminance factors of rock samples collected at the points photometered from the air. Such a check can be made with a field photoelectric spectrometer. The model of this instrument which we used in our work is described below.

The Field Photoelectric Spectrometer

The field spectrometer constructed in the Laboratory of Aeromethods in 1959 registers spectra photoelectrically and has a needle microammeter as an indicating device. This instrument is designed for ground measurements of spectral luminance factors in the wavelength range 400-1000 mμ. The construction of the instrument is briefly described in the papers of Voronkova et al. (1960) and Alekseev and Belov (1960).

The field photoelectric spectrometer has been extensively used in our work for control photometry of sand samples collected during an aeropetrographic survey of the Central Kara-kum.

A block diagram of the field photoelectric spectrometer is shown in Fig. 51. The optical unit I consists of a monochromator and a dispersing element — a diffraction grating; unit II is the radiation receiver, consisting of an FEU-22 photomultiplier; unit III is a d-c amplifier; unit IV is a 100-mA needle microammeter for displaying the signals taken from the load resistor of the photomultiplier.

The spectrometer is set for different wavelengths within the working range of the instrument by rotation of the diffraction grating. The resolution of the field photoelectric spectrometer is 10 mμ and the viewing angle is

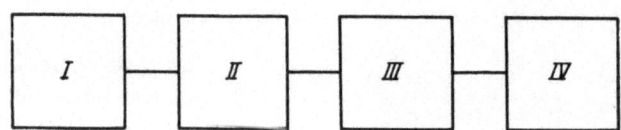

Fig. 51. Block diagram of field photoelectric spectrometer.

2°. The absolute mean square error of one measurement is 1%. The spectral width of the slit is about 8 mμ. The spectrometer has five stages of amplification of the sensitivity and is adapted for work on a tower or tripod (Fig. 52). The spectrometer is supplied by dry cells of the 315 A-E MGTs type. The set of four cells provides for operation of the instrument for 250-300 hr. The total weight of the instrument is 25 kg.

Photometry of the sample is best performed by two operators. One attends to the changing of the wavelength in the optical unit, the mounting of the sample in front of the monochromator entrance slit, and the correct sequence of photometry of the standard and sample. The second operator takes the readings of the microammeter and records the values in the field book. Photometry of a sample consists of the cycle: a) standard; b) sample; c) standard.

a) In photometry of the barite paper (standard) the sensitivity of the instrument is adjusted to show 100 divisions of the microammeter scale. This is effected by rotation of the regulator by the knobs of the step and smooth amplifier.

b) In the photometry of the sample the regulator is left in the same position as in the photometry of barite paper. The reading on the microammeter scale is the spectral luminance factor expressed as a percentage, since the microammeter scale is calibrated to give the measured values as a percentage of the signal maximum (taken as 100%) obtained in photometry of the standard.

c) In the repeat photometry of the standard the microammeter needle must again indicate 100%. If the needle deviates from 100% the measurement must be repeated again in the same order: standard, sample, standard.

The whole process of measurement of ρ_λ of one rock sample at 20-mμ intervals in the wavelength range 400 to 1000 mμ by two operators takes 7-10 min.

Processing of the data of measurements of reflected radiant energy in work with a field photoelectric spectrometer is less laborious than with other instruments. Moreover, the standard is measured before and after measurement of the sample. This dual check minimizes the error due to a change in light intensity in photometry of the sample and standard, i.e., ensures the most accurate values of ρ_λ.

Thus, in a summing-up of the characteristics of spectrometers employing photoelectric registration the following points can be noted:

1. In the discussed spectrometers the combination of diffraction grating and photomultiplier provides

Fig. 52. Field photoelectric spectrometer. Photometry of standard. Photo by K.E. Meleshko.

for spectrophotometry of objects in a wide region of the spectrum — from the ultraviolet to the infrared.

2. The use of photoemissive cells, which have a fast response, as detecting devices permits rapid registration of weak light flashes and, in view of this, spectrophotometry of small areas of surface can be carried out from an airplane in flight.

3. The use of a cathode-ray tube as a registering system avoids the use of the methods of photographic photometry; i.e., spectrometry can be performed more rapidly and with smaller errors.

4. Recording of the curves of distribution of reflected radiant energy on film or the use of a microammeter as an indicator does not stand in the way of the introduction of other improved registering systems involving computers.

The listed points favorably distinguish spectrometers employing photoelectric registration of spectra from spectrometers in which reflection spectra are registered photographically.

These brief descriptions of instruments for measurements of the spectral luminance of objects should be adequate for an understanding of the special features of the technique of measuring the reflecting properties of rocks.

For the correct registration of the values of the spectral luminance factors of investigated deposits and the geological interpretation of the values of ρ_λ, we will now have to describe the conditions of spectrophotometry of the objects and the methods of processing the results so that the obtained spectral luminance factors of the investigated objects can be compared with one another and with the composition of the investigated rock.

CHAPTER IV

CONDITIONS OF SPECTROPHOTOMETRY OF GEOLOGICAL OBJECTS. PROCESSING OF SPECTROGRAMS AND OSCILLOGRAMS

In addition to the conditions listed in the section dealing with the results of the experimental measurements of ρ_λ, we will have to consider the other circumstances and factors which can effect the accuracy of the values of ρ_λ obtained by aerial photometry.

Such factors affecting the value of ρ_λ of an investigated object include:

1) The height at which spectrophotometry is conducted. This height determines the size of the measured area and, hence, the homogeneity of the composition of the investigated rock.

2) The conditions of illumination at the instants of photometry of standard and object.

3) The types of films used for registration of the spectrograms and oscillograms, and the photography conditions during spectrophotometry.

4) The photographic processing of the exposed films.

5) The methods of measurement of the spectrograms and oscillograms.

Spectrophotometry of rock outcrops is carried out at particular points in the locality in accordance with previously charted itineraries. This is essential to secure a uniform distribution of the observed points within the investigated territory and to link these points with the topography of the region.

The choice of object is determined by the geologist, who is on board the airplane (helicopter) and visually observes the nature of the outcrops. The object designated for spectrophotometry is indicated to the pilot, who ensures the accurate passage of the apparatus over the object. On passage over the investigated object, a signal is given to the operatore to press the button of the cinecamera of the spectrograph (spectrovisor) and the time of photography is noted. The photometered point is recorded in the flight book under a particular sequence number and the conditions of photography are also noted [flight altitude at time of photometry, duration of photography, number of frames per second, width of spectrograph slit, state of light (presence of clouds and dust haze), type of film and filters used] and the photometered object is described. If necessary, the aircraft makes a landing and a rock sample is taken from the photometered area.

Height of Flight during Spectrophotometry of Geological Objects

The height of flight during the photometry of geological objects depends on the nature of the investigated rock outcrops. The more homogeneous their composition and structure, the greater the height from which they can be photometered. The areas of rock outcrop with homogeneous composition are usually relatively small and, hence, measurements of ρ_λ of a rock of particular petrographic composition necessitate photometry from low heights.

Experience has shown that a spectrophotometric survey is best conducted from a height of 10-50 m. This ensures that the photometered area is small (see Tables 5 and 6) and, hence, to some extend ensures that the rock within the measured area is homogeneous in petrographic composition and structure.

In photometry from a low height it is possible to select areas of outcrops with a relatively smooth and horizontal surface, i.e, to obtain even, unbanded spectrograms (see Fig. 25). A great advantage of spectrometry from low heights is that the thickness of the interlayer of air haze is insignificant and all the corrections for the effect of this haze on the results of measurements can be neglected.

Photometry from a low height in survey work on light airplanes of the YaK-12 type can be secured by instructing the pilot beforehand to make a low-level flight over the area designated for measurement. A spectrophotometric survey is much more easily conducted by a helicopter, which can pass over the object at low speed and, where necessary, land for the collection of samples at any spot.

Light Conditions and Choice of Time of Photometry

In the second chapter we showed that the nature or type of the indicatrix of diffusion of an investigated surface depends on the angle of incidence i of the light flux on the measured area and on the angle ε at which the reflected radiant flux is registered.

In measurement of the spectral luminance of an object from an airplane (helicopter) the instrument is mounted so that the optic axis of the spectrometer is perpendicular to the measured surface. In this case the angle ε between the direction of the beam reflected from the photometered area and the optic axis of the spectrometer is close to zero, the trigonometric factor in the luminance formula tends to unity, and, hence, in this case the size of the angle ε has no effect on the value of ρ_λ. The values of the measured ρ_λ with all other conditions of photometry equal will depend on the angle of incidence of the flux on the measured area and on the nature and properties of the investigated object.

The source of radiant energy in spectrophotometry in the field is the sun. Thus, the size of the angle i is determined by the time of day or by the angle of the sun, equal to $90 - h_\odot$. The effect of the angle i is greater the greater the deviation of the indicatrix of diffusion of the investigated object from the orthotropic type.

In our case the investigated objects (sands) have indicatrices of diffusion approximating those of orthotropically or diffusely reflecting bodies (see Figs. 3-5). The specular factor takes effect at angles i exceeding 70°.

The standard (barite paper) used in field spectrophotometry corresponds to an orthotropically reflecting body only at angles i in the range 0-45°. With increase in the angle i the specular factor begins to take effect and reaches a maximum at angles i equal to 70-80° (see Figs. 9-12).

Large values of angle i are observed in the morning or in the evening hours, varying with the geographical latitude of the locality and the season. The special tables given by Sharonov (1945) for calculation of natural illumination permit a rapid and easy calculation of the size of angle i in relation to the time of day for different geographical latitudes and different seasons. As an example, Table 7 gives the values, calculated from Sharonov's tables (see Appendix I), of the angles of height of the sun for the region of Darvaza (40° N, 59° 50' E; Central Kara-kum).

According to these tables, the size of angle i in the spring and summer months in the Darvaza region will

TABLE 7. Angles of Height of Sun $h_\odot$ (Degrees) for Darvaza Region

Date	Local Mean Time							
	9:10 AM	10:10 AM	11:10 AM	12:10 PM	1:10 PM	2:10 PM	3:10 PM	4:10 PM
5 May	34	45	55	63	67	62	53°30'	43°30'
11 May	35	46	56	64	68	64	54	44
16 May	35°30'	46°30'	57	65	69	64	55°30'	45
21 May	36	47	58	66	70	65	57	46
26 May	36°30'	47°30'	58°30'	67	70°30'	66	57°30'	46°30'
31 May	37	48	59	68	71	67	58	47
10 June	38	49	60	69	72	69	59	48
20 June	37	49	60	69	73	69	60	49
30 June	37	48	59	68	73	69	60	49

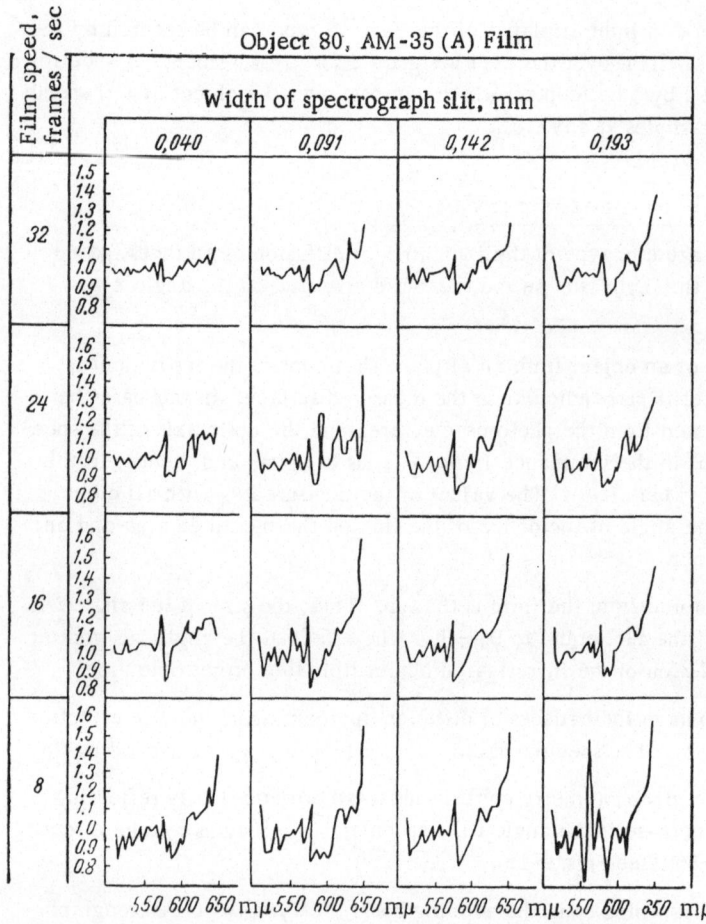

Fig. 53. Curves of relative transmission of two adjacent lines in relation to spectrograph slit width and film speed (frames/sec) for AM-35 cinefilm.

not affect the results of photometry between 10 AM to 4 PM local time. During this period the angles of incidence of the sun's rays do not exceed 45° from the zenith. This ensures the orthotropy of the indicatrix of our investigated objects and of the barite paper used as a standard.

Photometry of geological objects is best conducted in the first half of the day in view of: a) the change in spectral composition of the sun's light during the day; b) the cleaner and more transparent air in the first half of the day; c) in work in desert regions in the second half of the day, when air whirls and sandspouts due to heating of the sands by the sun are very common, a very thick unstable dust haze is formed and prevents photometry. Moreover, after midday, low-level flying becomes dangerous owing to the very strong ascending currents of air. We should also note that the general state of meteorological conditions is usually more stable in the first half of the day.

In view of the above and on the basis of many years' experience of aeropetrographic surveying, the most suitable time for photometry of geological objects within 39-42° N is the period from 10 AM to 2 PM local time.

Choice of Films for Registration of Spectra and Oscillograms

Photographic registration of the reflection spectrum of an investigated object depends on the correct choice of type of film, the conditions of photography, and the processing of the exposed films.

Before the type of film most suitable for registration of reflection can be chosen the reflecting properties of the objects for photometry must be known. For this purpose samples of the commonest types of rock in the region are photometered in the laboratory and the portions of the spectral luminance curves with the characteristic bends are determined. When the reflecting properties of the commonest rocks which are to be spectrophotometered in the field are known, the appropriate types of film can be chosen. The films are chosen by reference to the special manuals (Gorokhovskii and Barteneva, 1942; The Sensitometric Handbook, 1955; and others).

Preliminary observations of the spectral curves of sands of different petrographic composition showed that the main changes in the nature of the curves of ρ_λ occur in the region 500-650 mμ. Thus, from the available cinefilms we could select the most suitable for the registration of spectra in this wavelength region. In this connection we tested the following cinefilms in the Kizlyar region (northwest Caspian region) on August 4, 1957:

1) type AM-35 (A), produced in March 1957, emulsion 5590, reel 300;
2) type D, produced in May 1957, emulsion 450, reel 5198;
3) type MZ, produced in February 1957, emulsion 3800, reel 231.

TABLE 8. Relative Transmission of Two Adjacent Spectral Lines in Relation to Wavelength, Width of Spectrograph Slit, and Film Speed. AM-35(A) Film

Wavelength, mμ	0.040				0.091				0.142				0.193			
	8	16	24	32	8	16	24	32	8	16	24	32	8	16	24	32
510	1.096	1.000	1.014	0.998	1.137	1.066	1.074	1.014	1.074	1.023	1.045	1.057	1.109	1.101	1.089	1.084
515	0.847	0.946	0.938	0.968	0.867	0.823	0.897	0.971	0.853	0.906	0.902	0.929	0.849	0.904	0.910	0.900
520	0.968	0.940	0.993	0.989	0.883	0.948	0.946	0.989	0.899	0.906	0.910	0.968	0.904	0.873	0.883	0.912
525	0.885	0.927	0.951	0.951	0.889	0.904	0.931	0.971	0.906	0.931	0.916	0.929	0.900	0.940	0.966	0.948
530	0.986	1.021	0.998	1.007	1.019	0.982	0.971	0.966	0.993	0.982	0.986	0.993	0.986	0.957	0.989	0.954
535	0.916	0.940	0.951	0.973	0.895	0.873	0.935	0.962	0.920	0.908	0.908	0.929	0.875	0.916	0.920	0.929
540	0.916	0.946	0.935	1.000	0.953	0.959	0.957	0.942	0.938	0.942	0.942	0.959	0.916	0.948	0.912	0.940
545	1.014	0.968	0.968	0.986	0.971	0.948	0.989	0.982	1.023	1.044	1.002	0.980	1.030	0.971	1.000	1.002
550	0.931	0.980	1.007	0.986	0.971	0.910	0.951	1.007	0.933	0.935	0.904	0.986	0.971	0.966	0.942	0.966
555	1.000	0.998	1.002	1.009	1.007	1.091	1.030	1.012	1.018	1.009	1.000	0.986	1.000	1.000	1.018	0.993
560	0.977	0.993	0.986	0.971	0.966	0.942	0.916	0.966	0.971	0.957	0.997	0.998	0.681	0.959	0.959	0.971
565	1.023	1.002	0.998	1.028	1.061	1.000	1.028	1.007	1.007	1.030	0.989	1.007	1.446	1.023	0.993	0.977
570	0.893	0.923	0.990	0.935	0.904	0.879	0.910	0.929	0.859	0.889	0.879	0.912	0.875	0.900	0.929	0.938
575	1.054	1.137	1.124	1.074	1.132	1.172	1.119	1.124	1.145	1.135	1.132	1.076	1.109	1.106	1.064	1.099
580	0.873	0.796	0.883	0.927	0.843	0.767	0.871	0.861	0.800	0.809	0.801	0.879	1.000	0.818	0.839	0.849
585	0.904	0.862	0.912	0.942	0.869	0.863	0.867	0.918	0.826	0.847	0.845	0.904	0.733	0.873	0.883	0.873
590	0.853	0.881	0.891	0.953	0.857	0.838	1.241	0.906	0.877	0.873	0.853	0.897	0.863	0.843	0.863	0.883
595	0.959	0.986	0.968	0.953	0.971	0.942	0.948	0.982	0.916	0.938	0.910	0.912	1.000	0.959	0.977	0.948
600	0.948	0.982	1.000	1.002	0.986	0.986	0.971	0.986	0.957	0.989	1.007	1.000	1.000	0.982	0.938	0.942
605	0.989	1.000	0.948	0.977	0.957	0.957	0.971	1.000	0.986	1.007	0.953	0.982	0.942	1.007	0.959	1.023
610	1.074	1.057	1.044	1.049	1.091	1.014	1.035	1.064	1.124	1.061	1.111	1.045	1.062	1.074	1.138	1.079
615	1.054	1.119	1.109	1.049	1.111	1.091	1.180	1.074	1.091	1.074	1.106	1.076	0.863	1.079	1.035	1.062
620	1.091	1.052	1.018	1.118	1.054	1.076	1.042	1.014	1.061	1.047	1.042	1.018	1.030	1.086	1.125	1.074
625	1.042	1.023	1.048	1.018	1.049	1.091	1.018	0.916	1.091	1.047	1.047	1.064	1.062	1.030	1.035	1.007
630	1.079	1.074	1.137	1.044	1.124	1.137	1.099	1.186	1.137	1.091	1.079	1.076	1.125	1.106	1.099	1.102
635	1.111	1.061	1.096	1.040	1.091	1.145	1.084	1.079	1.137	1.125	1.109	1.089	1.178	1.149	1.140	1.109
640	1.286	1.137	1.135	1.064	1.208	1.208	1.153	1.079	1.236	1.164	1.233	1.091	1.247	1.172	1.172	1.167
645	1.124	1.135	1.052	1.014	1.221	1.230	1.993	1.061	1.259	1.250	1.273	1.119	1.267	1.233	1.259	1.225
650	1.403	1.042	1.183	1.101	1.549	1.596	1.432	1.267	1.545	1.489	1.406	1.241	1.633	1.426	1.368	1.399

Slit width, mm

Film speed, frames/sec

TABLE 9. Relative Transmission of Two Adjacent Spectral Lines in Relation to Wavelength, Width of Spectrographic Slit, and Film Speed. D Films

Wave-length, mμ	Slit width, mm															
	0.040				0.091				0.142				0.193			
	Film speed, frames/sec															
	8	16	24	32	8	16	24	32	8	16	24	32	8	16	24	32
510	1.279	1.122	1.140	1.099	1.132	1.132	1.104	1.112	1.086	1.170	1.159	1.086	1.064	1.170	1.227	1.132
515	0.798	0.774	0.855	0.900	1.000	0.849	0.984	0.863	0.883	0.869	0.834	0.953	0.959	0.935	0.834	0.906
520	1.064	1.170	1.028	1.021	1.000	1.069	1.059	1.050	1.132	0.959	0.975	1.026	1.086	0.929	1.000	1.016
525	1.000	0.991	0.933	0.883	0.959	0.863	0.914	0.929	0.959	1.042	1.132	0.953	1.000	1.026	0.984	1.042
530	1.154	1.059	1.170	1.072	1.064	1.279	1.003	1.009	0.959	1.042	0.959	1.026	0.979	1.016	1.086	0.929
535	0.782	0.828	0.873	0.914	0.849	0.828	0.891	0.991	0.920	0.906	0.920	0.944	0.902	0.920	0.914	0.953
540	1.000	0.900	0.873	0.863	1.021	0.914	1.000	0.877	0.920	0.918	0.883	0.900	1.021	0.849	0.869	0.891
545	1.086	1.009	1.000	1.050	1.086	0.891	0.914	0.935	0.959	0.867	0.953	0.959	0.883	1.000	0.940	0.953
550	0.815	0.869	0.855	0.877	0.750	0.968	0.935	0.953	0.920	0.959	1.009	0.906	1.000	0.920	0.940	0.849
555	0.902	1.009	1.033	0.906	1.109	0.991	0.975	1.016	1.042	0.959	0.891	0.991	0.959	0.959	0.959	1.000
560	1.000	0.900	0.906	1.009	0.798	0.953	0.953	0.906	0.883	0.883	0.840	0.944	1.000	0.959	0.920	0.920
565	1.109	0.968	0.984	1.033	1.202	1.042	1.009	1.026	1.000	1.042	1.132	0.975	1.000	1.042	1.000	1.009
570	0.867	0.869	0.877	0.849	0.815	0.849	0.900	0.869	0.920	0.883	0.920	0.900	0.883	0.832	0.920	0.935
575	0.959	1.271	1.159	1.170	1.064	1.247	1.033	1.279	1.086	1.178	1.059	1.227	1.086	1.064	1.042	1.140
580	1.132	0.787	0.841	0.869	0.959	0.869	0.953	0.815	0.920	1.000	1.009	0.828	0.883	0.867	1.042	0.849
585	1.154	1.159	1.059	1.009	1.021	0.920	1.170	1.016	0.920	0.902	0.841	1.016	0.920	0.940	0.902	0.959
590	0.832	0.828	0.820	0.953	0.902	1.000	1.059	0.944	0.959	0.867	1.050	0.841	1.000	0.940	1.000	0.902
595	1.042	1.016	1.132	1.026	0.959	0.984	0.953	0.929	0.959	1.132	1.000	1.042	1.042	0.959	0.902	1.021
600	1.064	1.094	0.891	0.935	1.021	1.042	1.077	0.953	1.086	0.920	0.968	0.959	1.000	1.086	1.000	1.000
605	1.000	0.877	0.968	0.944	0.832	0.881	0.929	0.959	0.849	0.940	1.000	0.944	0.959	0.883	0.940	0.867
610	0.849	1.086	0.991	1.016	0.959	1.047	0.877	1.026	1.000	0.959	0.910	1.009	0.920	1.021	1.064	0.940
615	0.940	0.900	0.975	0.929	0.883	0.975	0.900	0.959	1.000	1.000	0.979	1.026	1.042	0.959	0.979	0.883
620	0.959	0.906	1.009	0.935	0.959	0.883	1.159	0.935	1.042	0.920	0.920	0.883	0.883	0.940	0.940	0.920
625	0.902	0.920	0.929	0.953	0.959	1.086	0.906	0.953	0.920	0.979	0.920	0.935	1.000	0.867	0.893	1.021
630	1.154	1.132	1.122	1.170	1.132	0.940	1.159	1.104	1.042	1.178	1.042	1.151	1.086	1.154	1.154	1.086
635	1.202	1.279	1.377	1.279	1.086	1.419	1.346	1.400	1.279	1.306	1.600	1.346	1.178	1.361	1.361	1.396
640	1.932	2.080	1.733	1.914	2.183	2.028	1.853	1.726	1.671	1.862	1.750	1.866	1.706	1.853	1.778	1.884
645	2.014	1.758	1.722	1.528	2.094	1.905	1.963	1.799	2.051	2.239	2.203	1.862	2.014	2.334	2.014	2.148
650	3.013	2.080	1.742	1.837	2.911	2.541	2.203	1.762	3.236	3.013	2.432	2.080	2.985	2.553	2.891	2.512

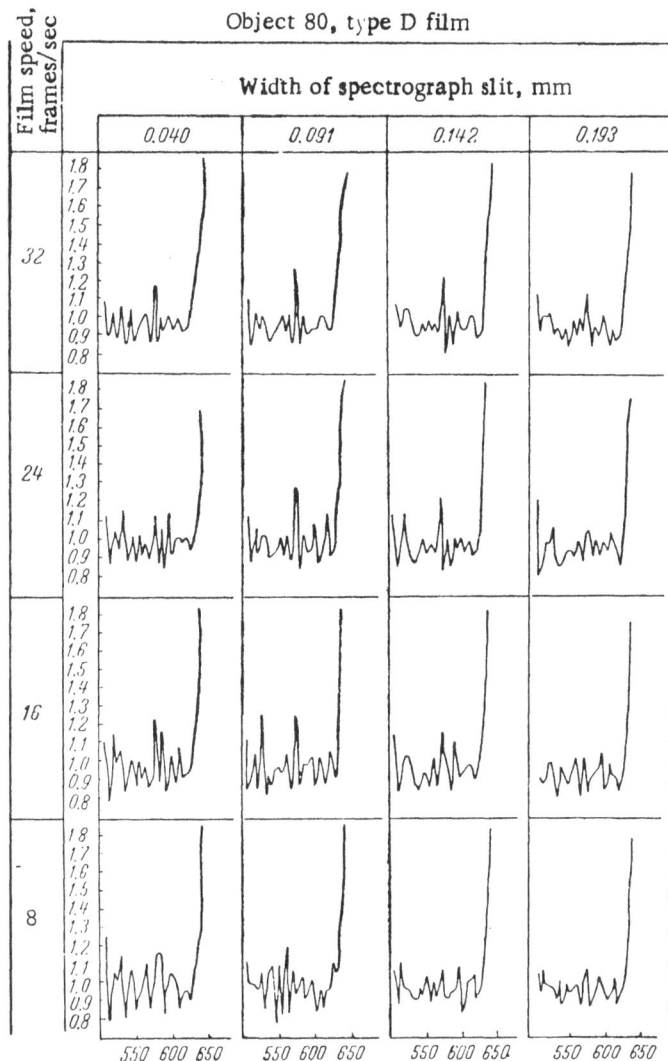

Object 80, type D film

Width of spectrograph slit, mm

Film speed, frames/sec

Fig. 54. Curves of relative transmission of two adjacent lines in relation to width of spectrograph slit and film speed (frames per sec) for type D film.

The tests of the films consisted in repeated photometry of the same region of an experimental polygon with four different exposures and four different widths of the spectrograph slit. In every case photometry was performed from the same height – 15-20 m – under a clear, cloudless sky. The tests took 13-14 min, so that the slight change in the angle of incidence of the sun's rays could not affect the nature of the indicatrices of reflection of the object or standard.

In the analysis of the spectrograms, we measured the relative transmission of two adjacent spectral lines in relation to the wavelength. The results of measurements of the spectrograms obtained in the photometry of predominantly quartz sands of the Volga type on AM-35 (A) film and on D film are given in Table 8 and Table 9, respectively. The variation of the shape of the curve of the relative transmission of two adjacent lines with the width of the spectrograph slit and the film speed (frames/sec) is depicted in the form of tables (Figs. 53 and 54). These tables show that with AM-35 film the photometry of sands is best performed with a slit width of 0.142 mm and a film speed of 16 or 24 frames/sec. When the spectrograph slit is less than 0.1 mm wide, the film speed must be reduced to 16 frames per sec. In the case of a wide slit (0.193 mm) and a low film speed, the obtained spectrograms have high density and their measurement is difficult. In the case of a narrow spectrograph slit (0.040 mm) and a high film speed, the registered spectrograms have a very low density and are unsuitable for measurement.

If the spectrum is registered on D film, the measurement of the spectrograms is made difficult by the large size of the emulsion grains. This is indicated by the dispersion of the curves of relative transmission of two adjacent lines (Fig. 54). In the case of registration of the spectrum on D film, photometry can be carried out with narrow spectrograph slits and at high film speeds (32 frames/sec). If the spectrograph slit width is fixed and low film speeds are used, the spectrograms obtained on D film are very dense and their measurement is almost impossible.

Registration of spectra on MZ film showed that the density of the obtained spectrograms is inadequate and their measurement is difficult. But spectrograms obtained on MZ film with a wide spectrograph slit (0.193 mm) and at a low film speed (8 frames/sec) have satisfactory density. With a medium or narrow spectrograph slit and a relatively high film speed, the spectrograms obtained on MZ film are very thin and cannot be measured.

For the obtention of reliable values of ρ_λ of an investigated object it must be photometered under exactly the same conditions as the standard in relation to which its relative spectral luminances are measured. A spectrogram for measurement with a microphotometer is obtained from the photometered standard by an appropriate choice of width of spectrograph slit and film speed. In this case the selected slit widths and frame rates must remain the same in the photometry of all the investigated objects which are to be compared with one another.

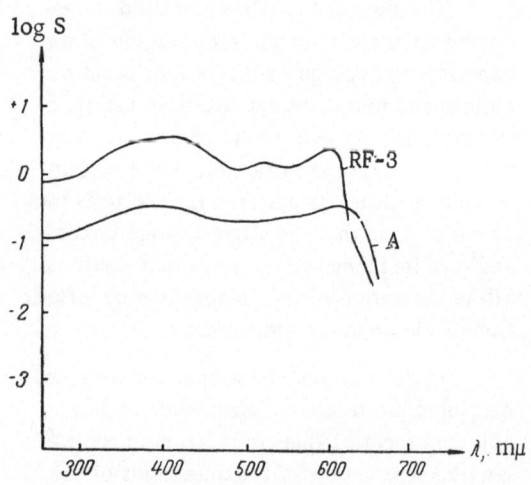

Fig. 55. Curves of spectral sensitivity of films of
type AM-35 (A) and RF-3.

For aerial spectrophotometry of sands on extensive areas of desert we used AM-35 film. The photometry was performed with a slit width of 0.142 mm and a film speed of 24 frames/sec. The AM-35 panchromatic cinefilm chosen for registration of the spectra has medium sensitivity and normal contrast. The spectral sensitivity curve is illustrated in Fig. 55, which shows that the sensitivity of AM-35 film is more or less constant in the visible region of the spectrum and falls away sharply in the 650-mμ region. Thus, within the visible region of the spectrum, registration of reflection spectra on AM-35 film is sufficiently reliable.

Besides registration of the reflection spectrum the same film is used to photograph the external appearance of the investigated object. The main difficulty encountered in this case is the blurring of the photograph which inevitably occurs in photometry from an airplane (from heights of less than 50 m). For the prevention of blurring, the exposure must be as short as possible. In standard cinecameras the minimum exposure is about 0.01 sec. A further reduction in exposure can be secured by a reduction of the slits in the shutter. Accordingly, the slit in the shutter in the spectrovisor was reduced to a size giving an exposure of 1/300 sec, which provides a fairly distinct photograph of the photometered area (see Fig. 49).

An effective way of increasing the definition of the image in some cases is by the use of some kind of filter, the choice of which is based on the same considerations as in ordinary photographic practice. Filters of glass tinted in bulk, which have reliable optical characteristics (Catalog of Colored Glass), are most durable in expeditionary conditions. The filter fitted on the objective of the landscape system is chosen so that the best photograph of the area of rock is obtained. It is obvious, of course, that no filters of any kind should be fitted on the spectrograph objective.

The introduction of an optical wedge is not recommended for the photometry of areas of rock, since the diffraction, even though slight, caused by this device affects the nature of the spectrogram.

Thus, when the most suitable of the available types of film for registration of the reflection spectra and the best conditions of recording the spectrograms have been chosen, a high-quality photograph of the investigated rock surface can be obtained on the same film by a suitable choice of filter.

The selected conditions of filming during the photometry of standard and objects must be kept the same in all compared cases.

In the case of cinematographic recording of curves from the screen of a cathode-ray tube, the curves are photographed on film with high sensitivity and high contrast. In practical work with the spectrovisor, film of the RF-3 type was used. This is an orthochromatic film with high sensitivity and high contrast (Handbook of Cinematographic Technique, 1960). The spectral sensitivity of RF-3 film is illustrated in Fig. 55, which shows that the film is sensitive in the green part of the spectrum. This ensures registration of the curves from the cathode-ray tube screen, which have a green color.

In view of the fact that a considerable number of the discarded measurements of ρ_λ (in aerial photometry of geological objects with spectrometers employing photographic or photoelectric registration of reflected energy) are due to errors in processing the exposed films, we will dwell a little more fully below on the processing of films in the field.

Processing of Exposed Films

Before beginning trial developments of the exposed films, we had to refer to the existing manuals and select a general processing procedure for the film of the type used in aerial photometry.

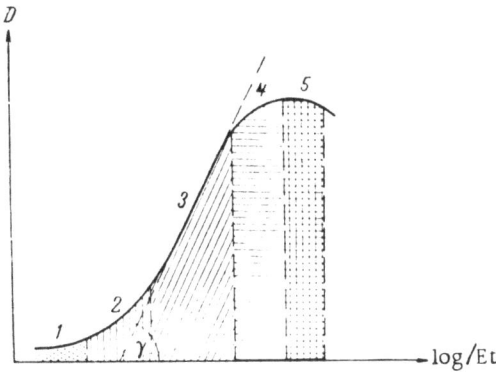

Fig. 56. Schematized characteristic curve.

The main problem in the processing of the exposed film lies in the choice of the optimal development procedure, giving preference to the quality of the photograph of the spectrum, rather than to the photograph of the locality with the investigated area. Once this optimal procedure has been chosen it must be used for the processing of all the films exposed during photometry on the investigated area.

The technique of photographic photometry is based on measurement of the optical densities of the spectrograms registered on the film. The optical density D is the logarithm to base ten of the reciprocal of the transmissivity T. The transmissivity of the measured area of film is defined as the ratio of the amount of light J transmitted by the blackened part to the amount of light J_0 transmitted by the clear part:

$$T = \frac{J}{J_0}.$$

The optical density depends not only on the effect of the light flux on the light-sensitive layer of the film, but also on the conditions of processing of the exposed film (on the composition and temperature of the developer, the development time, the thickness of the emulsion). Hence, it is very important to follow the correct procedure in the processing of the exposed films.

During development of the exposed film, the silver reduced in the sensitive layer produces blackening of various densities, which is related to the total amount of radiant energy of a given wavelength E_λ incident on the film during a time t. This relationship is complex, changing at different stages of action of the radiant energy, and it is usually represented by a graph. The curve representing the relationship between the density D and the logarithm to base ten of the intensity E during its action on the film (in seconds) during photography (log Et) is called the c h a r a c t e r i s t i c   c u r v e. In practical spectral analysis the logarithm to base ten of the light intensity (log J) is often plotted on the x axis instead of the logarithm of the amount of light.

Figure 56 shows a schematic characteristic curve, where the regions corresponding to different relationships between D and log Et are shaded conventionally. In the region of fogging, 1, of the film (plate), D does not depend on the conditions of development of the film but depends on the quality of the emulsion; in the underexposure region, 2, the density varies very slightly and its measurement does not give reliable results; in the normal-exposure region, 3, the density varies almost linearly with the logarithm of the light intensity; in the overexposure region, 4, the variation in the density is again nonlinear and measurement of D in this region cannot give reliable results; in the solarization region, 5, there is an inverse nonlinear relationship, and this region is totally unsuitable for photometric measurements.

The angle γ made by the straight-line portion of the characteristic curve with the x axis determines the contrast of the emulsion; the greater this angle, the higher the contrast of the emulsion.

Aerial spectrographs and spectrovisors for spectrophotometry use 35-mm film. The standard magazine of cinecameras is designed for 30 m of film. This length of film is sufficient for the spectrophotometry of 40-50 objects, i.e., one magazine of film is sufficient for a whole working day, and the spare loaded magazine is usually brought back unused from the flight.

For a check of the quality of the photometry, the exposed film must be processed in the field photographic laboratory immediately after return from the trip.

Practical field work has shown that the development of a 30-m roll of film is conveniently performed in a vinyl plastic tank with a diameter of 45 cm, a height of 20 cm, and a capacity of about 30 liters; the amount of developer is 15 liters. The exposed film is wound on a special spiral holder made of stainless steel. The total diameter of the spiral is 43 cm, the width between the ribs is 0.5 cm, and the height of the ribs is 4 mm. In the center of the spiral there is a spool with a slot, into which one of the ends of the exposed film is inserted and fixed (it must be remembered that if the film is stretched it may jump out of the grooves of the spiral, and then rejects due to sticking on different parts of the film are unavoidable). When the film has been wound on the spiral, the latter is immersed in the developer solution. Here some precautions must be observed – the spiral with the film must be raised and lowered several times when it is immersed in the solution. This ensures the removal of air bubbles, which stick to the emulsion.

During the development of the film, the spiral holder is periodically turned in the direction in which the film is wound. This agitation is essential to secure uniform development throughout the length of the film.

The exposed film is best developed in a normal fine-grain developer. In field work, satisfactory results were given by compensating metol-hydroquinone developers D-76 and K.V. Chibisov developer, which ensures a great range of density of the processed material with a fine-grained image. These developers gave spectrograms of the required quality even in conditions deviating considerably from the standard conditions. For instance, the films were processed in solutions at a temperature of about 22° C, instead of the generally adopted 18-20°. The mineral content of the water was high and in many cases Trilon B had to be used to purify the water. The time required for the development of 30 m of film in a plastic tank containing 15 liters of developer, at a temperature of + 22° C, was 8 min.

The film was regarded as properly developed if the contrast factor and general development in all parts of the film were the same.

In work in desert conditions, where the air temperature is much higher than desirable, the films must be fixed in an acid hardening bath to prevent spreading of the gelatin layer of the film, although in this case there is some deformation of the base due to the hardening.

Throughout the fixing of the film it is essential to secure uniform mixing of the solution. Fixing is stopped when the emulsion on the film becomes transparent. In our work the time spent by the 30-m film in the fixing solution was 10-15 min.

After removal from the tank with the fixing solution the film is thoroughly washed in running water or repeated changes of water. The period of washing should not be less than 30 min.

The thoroughly washed film is wound on a special drum for drying and the excess water is removed from it with a cotton wad (the drying drum can be of any size – its shape and size usually depends on the material available).

In work with films in climatic conditions where the air temperatures are high a refrigerator should be an essential part of the equipment. If a refrigerator is not available, the solutions can be kept in damp wrappings.

The development of each individual film is checked by superimposing a sensitometric wedge at the beginning and end of the film. A sensitometric wedge is an optical wedge with steps of regularly increasing density. The sensitogram obtained by superimposing the wedge is required for plotting the characteristic curve of the emulsion used. Sensitometric standardization in field conditions is carried out from the light intensity scale on a field sensitometer. In the superimposing of the sensitometric wedge, an accurately standardized light source with known technical spectral specifications is used. In our case the source was an incandescent lamp operating off a ZnI-60 storage battery. The exposure for superimposing was 1/20 sec, which was secured by a leaf-type shutter with an iris diaphragm.

The diffuse density of the superimposed sensitogram was measured with a densitometer. (We used an IFT-11 or DEE-10 visual wedge densitometer.) From the densities obtained by measurement the characteristic curves of the film are plotted and from the shape of these curves the satisfactoriness of development of the exposed film is determined.

Development is regarded as satisfactory if the contrast factor γ or the gradient of the straight-line portion of the characteristic curve lies within the interval of values established for the measurements and is always the same for spectrograms compared with one another. In our field work the contrast factor of the developed films for one of the investigated regions was 0.65 ± 0.05, and in work in another region the films were developed to $\gamma = 0.70 - 0.71$.

The processing of RF-3 film exposed in spectrophotometry with a spectrovisor required developers giving greater contrast (more concentrated Chibisov developer or D-19 developer was used). Otherwise the general processing of RF-3 film was performed under exactly the same conditions as in the case of processing of AM-35 film.

Thus, by a correct choice of conditions of photography and processing of the film exposed in spectrophotometry from a helicopter(airplane) we obtain a series of films with spectrograms or oscillograms and photographs of the investigated areas. The next step is conversion to spectral luminance factors of the photometered objects by measurement of the obtained spectrograms and oscillograms.

Measurement of Spectrograms and Oscillograms. Plotting of Spectral Luminance Curves of Investigated Objects

In aerial spectrophotometry the luminance properties of the investigated rock are registered in different ways according to the instrument used to measure ρ_λ. The records consist either of photographs of the reflection spectrum (spectrograms) or photographs of the curves of spectral distribution of the reflected energy (oscillograms). The processing of spectrograms and oscillograms is quite different, since in one case the densities of the film have to be measured, while in the other case only the ordinates of the curves are measured. In view of this, separate descriptions of the methods of measuring spectrograms and oscillograms are given.

Measurement of Spectrograms and Plotting of ρ_λ Curves

As a result of aerial photometry of rock outcrops with the aerial cinespectrograph, the exposed film carries images of a series of spectrograms of the investigated objects and the standard, which is photographed through an optical wedge.

Thus, the main rule of photographic photometry – the obtention on the same film with equal exposures of two photometrically equal results [one from a weaker source (the object in our case) and the other from a stronger source (the standard in our case) weakened to a known degree] – is observed in this case.

The obtained spectrograms are converted to the spectral luminance factors of the photometered objects by measurement of the densities of these spectrograms on any kind of microphotometer. Photoelectric microphotometers have recently been widely adopted in laboratory practice. Such instruments are the MF-2 nonrecording photoelectric microphotometer (Krupp and Lerner, 1950) and the MF-4 recording photoelectric microphotometer (Kosov, 1960).

The essential feature of these microphotometers is that the light flux transmitted through the investigated part of the spectrogram falls on a photocell and excites a proportional photocurrent in the latter. The photocell current, delivered to a galvanometer, causes a rotation of the frame with the mirror, and the swing of the latter is projected onto a reading scale. The scale is calibrated in accordance with the sensitivity of these photometers, which can detect a change in photographic density of approximately 1/250.

Adequate accounts of methods of measuring spectrograms are given in various manuals, on one of which (Prokof'ev, 1951) the conduction of our work was based.

In measurement of the relative transmission intensity or optical density by photographic photometry, the basic principle is that from two unequal densities J_1 and J_2 it is possible to obtain equal densities by attenuation by factors p_1 and p_2. For radiations of equal wavelength the following condition is valid: $J_1/J_2 = p_1/p_2$.

The optical density from the standard is weakened by means of a special optical wedge. This device is a quartz or glass plate 0.5 mm thick, 7 mm wide, and 36-40 mm long, with a series of bands, usually seven to nine, of different transmissivity. These steps are formed by the deposition of evaporated platinum. There are opaque partitions between the steps. The steps are graduated in logarithms of the transmissivity. For convenience of calculation the transmissivity of the extreme steps is made equal to 100 and 10. Table 10 shows the graduation of the steps of the wedge for the UF-281 spectrograph. This was the one that we used in our work.

TABLE 10. Calibration of Optical Wedge Used in the Work

No. of step	Logarithm of transmissivity J
1	2.00
2	1.83
3	1.71
4	1.59
5	1.54
6	1.31
7	1.16
8	1.00
9	Transparent

The optical wedge is mounted either directly in front of the spectrograph slit or can be placed in the mount of the first lens of the intermediate-image system, and the second lens will project a reduced image of the wedge onto the spectrograph slit. In our arrangement the optical wedge was mounted directly in front of the spectrograph slit and could be removed if necessary. When the optical wedge is inserted, the spectrogram of the photometered standard is of the banded type (see Fig. 41).

The characteristic curve of the exposed and processed film is plotted from the data of density measurements on the spectrogram obtained by photometry of the standard with the optical wedge inserted. The densities are measured at fixed wavelengths for all nine steps of the spectrogram. Several characteristic curves are plotted for each film so that the whole length of the investigated spectral region is characterized. By plotting a series of characteristic curves we can select the straight-line, normal-exposure portions of these curves. In this case the projections of the ends of the straight-line portions of the curves on the y axis are taken and this fixes the extreme values of the densities.

When the limits of the straight-line portions of the characteristic curves of the exposed film have been found we can proceed to the measurement of the spectrograms obtained in the photometry of the investigated objects.

From the series of spectrograms of each investigated object on one film we select a few spectrograms (usually three) obtained by photometry of the cleanest and most homogeneous areas of the rock outcrops. The nature of the surface of the photometered area can be checked either on a spectrum projector or by visual examination of the negative. In the case of photometry of barchans the photograph shows a clean homogeneous field, usually covered with fine wind-ripple marks. The spectrogram of such an area has no longitudinal bands, such as are found in the case of photometry of an area which is rough or has other formations – shrubs, salt efflorescences, shaded spots.

For the obtention of the spectral luminance factors of the investigated object, the density D_s of the step of the standard closest to that of the spectrogram of the object is measured, and the density of the object D_{obj} is

TABLE 11. Scheme of Recording Measurements in the Processing of a Spectrogram on the MF-2 Microphotometer and Conversion to Spectral Luminance Factors

Wavelength, mμ	Density		Difference in logarithms of intensity Δ log J	Anti-logarithm ΔJ	ϱ_λ, %
408	61.0	70.0	0.180	1.514	75.7
421	59.0	71.0	0.095	1.235	61.7
444	61.0	74.4	0.110	1.290	64.5
459	62.0	78.3	0.130	1.354	67.7
472	64.0	78.8	0.120	1.318	65.9
485	64.0	80.0	0.135	1.368	68.4
498	64.0	80.0	0.135	1.368	68.4
511	64.0	84.4	0.160	1.445	72.2
524	68.0	88.5	0.170	1.478	73.9
537	72.0	91.5	0.160	1.445	72.2
550	75.0	93.0	0.150	1.413	70.6
563	78.5	94.0	0.130	1.354	67.7
576	79.0	93.5	0.125	1.334	66.7
589	79.0	93.5	0.125	1.334	66.7
602	79.0	93.0	0.120	1.318	65.9
615	73.5	87.0	0.110	1.290	64.5
628	72.0	82.0	0.090	1.230	61.5
641	66.0	72.0	0.060	1.135	56.7
654	48.0	43.0	0.040	1.085	54.2

measured. Measurements are made in each wavelength interval, equal for the standard and investigated object. The obtained densities are plotted on the characteristic curve and are projected from the curve onto the x axis, which gives the values of the logarithms of the transmissivity log J. The difference of these logarithms gives the ratio of the amount of light arriving from the reflecting surface of the object to the amount of light reflected from the standard and weakened by a known factor. The attenuation factor of the standard spectrogram step chosen for comparison is determined from the graduation of the steps of the optical wedge (Table 10).

To avoid errors in measurement of the spectrograms we can use only the normal-exposure portion, where the characteristic curve is linear and the density is given by the relationship

$$D = \gamma \left(\log J_1 - \log J_2 \right).$$

The stages of work with the MF-2 microphotometer are as follows:

1. The film with the spectrograms is fixed on the microphotometer stage so that it can easily be shifted parallel to the boundaries of the spectrogram steps.

2. The spectrogram of the standard photographed with the optical wedge inserted is chosen and the density of all the steps is measured. Knowing the transmissivity of the wedge's steps, we plot the characteristic curve from the measured densities.

3. Among the steps of the standard spectrogram we select the step with density close to the density of the photometered object so that we can work on the straight-line portion of this characteristic curve. The densities are measured over the spectrum at fixed wavelength intervals or distances. In thorough measurements the size of each step can be 10 or 20 mμ. For convenience in subsequent treatment of the spectral luminance curves, we divide the whole measured part of the spectrum by 19 points at equal wavelength intervals. The data of the measurements of the spectrogram are written in the column D_S (Table 11).

4. From the series of spectrograms obtained by photometry of the object we select three frames, the quality of which is guaranteed by the cleanness of the surface of the photographed outcrop. Each of the selected three frames is measured on the microphotometer. The step between the measurements at each wavelength is the

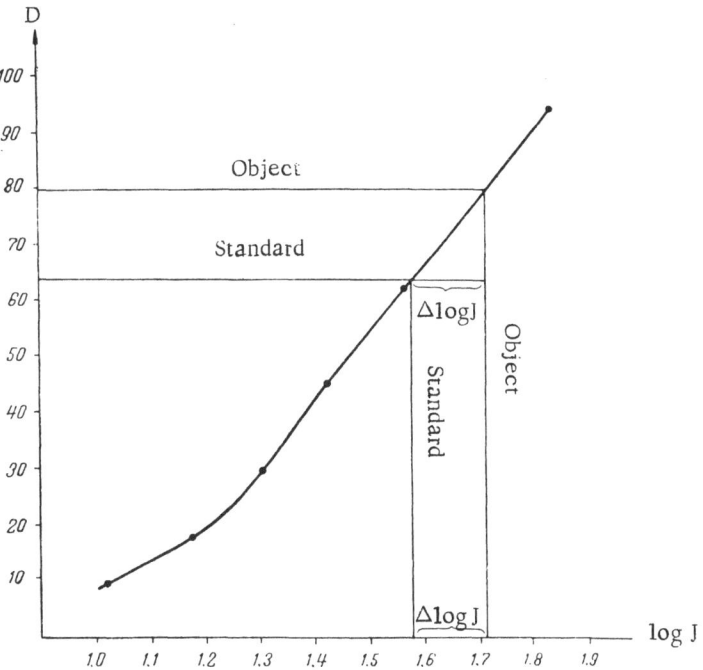

Fig. 57. Graph illustrating conversion from density D of measured spectrogram to transmission intensity log J of standard and object.

same as in measurement of the spectrogram of the standard. The mean values of the measured densities for the three frames are written in the column D_{obj} (Table 11).

5. Using the characteristic curve, plotted as instructed in stage 2, we find from the ordinates of the densities of the standard and object the corresponding points on the x axis, i.e., we find the difference in the logarithms of the transmission intensities $\Delta \log J$. Graphical conversion from density D to transmission intensity J is shown in Fig. 57.

6. The values of the logarithms are converted to antilogarithms and the obtained data are multiplied by the attenuation factor of the standard spectrogram step taken for comparison in the measurements.

The ratio of the transmission intensity of the object to the transmission intensity of the standard, expressed as a percentage, is the required spectral luminance factor. The spectral luminance curve of the investigated object is plotted from the obtained values of ρ_λ.

Table 11 shows an example of measurements of the densities of the spectrograms of the standard and spectrograms obtained in aerial photometry of the surface of graywacke sand from the region of the Sulak delta (northwest Caspian region). The fifth column of this table gives the antilogarithms of the transmission intensity. Multiplication of these values by the transmission factor of the wedge step (in this case log 1.71 or 0.5) gives ρ_λ.

The processing of the spectrograms can be automated by measurement of the densities on a recording microphotometer. For instance, with the MF-4 photoelectric recording microphotometer we can measure the optical densities and observe the readings without recording.

The operation of this microphotometer is based on registration of the variation of optical density of the spectrogram by means of a recording system. The light beam passing through the measured spectrogram falls on the photocell, thus giving rise to a current, which is transmitted to the galvanometer and causes a rotation of the frame with the mirror. The deflection of the mirror is proportional to the photocurrent and depends on the

TABLE 12. Example of Calculation of Spectral Luminance Factor of Sand Sample from Measurements of Density of Spectrum with MF-4 Microphotometer

Wavelength, mμ	$\log \dfrac{J_{obj} \cdot 16}{J_s}$	$\dfrac{J_{obj} \cdot 16}{J_s}$	ϱ_λ, %
510	0.204	0.246	24.6
515	0.193	0.240	24.0
520	0.200	0.244	24.4
525	0.206	0.247	24.7
530	0.228	0.260	26.0
535	0.265	0.284	28.4
540	0.240	0.268	26.8
545	0.261	0.271	27.1
550	0.256	0.278	27.8
555	0.258	0.279	27.9
560	0.269	0.286	28.6
565	0.264	0.283	28.3
570	0.283	0.295	29.5
575	0.275	0.290	29.0
580	0.326	0.326	32.6
585	0.296	0.303	30.3
590	0.302	0.308	30.8
595	0.302	0.308	30.8
600	0.279	0.293	29.3
605	0.249	0.273	27.3
610	0.260	0.280	28.0
615	0.283	0.295	29.5
620	0.306	0.311	31.1
625	0.294	0.303	30.3
630	0.302	0.308	30.8
635	0.284	0.296	29.6
640	0.290	0.300	30.0
645	0.288	0.299	29.9
650	0.326	0.327	32.7

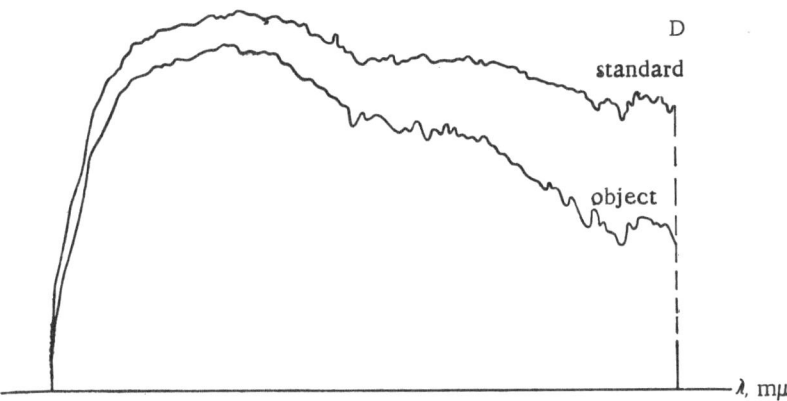

Fig. 58. Curves of densities of spectrograms of standard and object obtained by measurement of films on MF-4 recording microphotometer.

density of the measured part of the spectrogram. A view of curves obtained by measurement on the MF-4 is shown in Fig. 58.

The data of spectrogram measurements obtained with the recording microphotometer are subsequently treated by the method described above (Table 12).

All the measurements of the blackening marks and the investigated spectrograms must be carried out with the same setting of the microphotometer for zero deflection (transparent part of film). With this value constant the deflection of the galvanometer, corresponding to the measured spectrogram, uniquely determines the density.

It is obvious from the description of the methods of measuring spectrograms on microphotometers that this is a fairly long process, which is not devoid of errors. We need only mention the errors associated with the conditions of illumination, inhomogeneity of coating of the film, variations in the development of the films; errors associated with additional illumination from the jaws of the spectrograph slit (Podmoshenskii et al., 1960). This induced us to design instruments in which registration of the reflected energy is based on a different principle. Such instruments are the above-described spectrometers employing photoelectric registration of the reflection spectra.

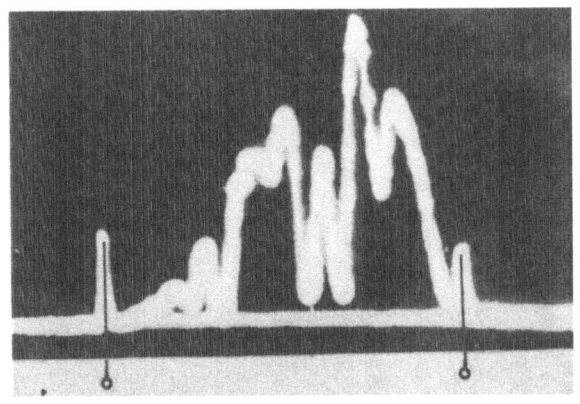

Fig. 60. The reference marks at the ends of the curves serve to determine the scale in wavelength division of the oscillogram.

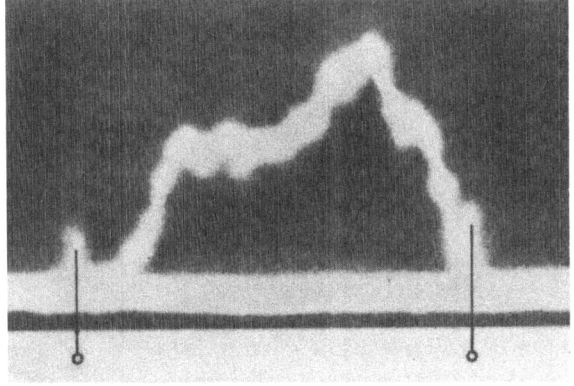

Fig. 59. Enlarged print of oscillogram obtained by spectrophotometry with the 1959 model spectrovisor.

Measurement of Oscillograms and Plotting of ρ_λ Curves

In the case of work with spectrovisors the film carries oscillograms — curves of the spectral distribution of the radiant energy reflected either from the standard or from the investigated object. For the obtention of the values of the spectral luminance factors the oscillograms are measured. The measuring procedure is as follows:

1. The oscillogram from the standard is traced on tracing paper. For this purpose an enlarger or spectrum projector is used. Enlarged prints of these oscillograms can also be obtained (Fig. 59).

2. In correspondence with the calibration curve the horizontal axis on the drawing or print of the oscillogram is marked off in wavelengths at selected intervals.

3. The oscillogram from the investigated object is traced on tracing paper or an enlarged print is obtained on photographic paper. It is essential to ensure that the scale and divisions of the wavelength axis are the same as in the case of the oscillograms of the standard. Preservation of the same scale in division into wavelength intervals is secured by the superposition of reference marks (Fig. 59 and Fig. 60). Special graticules can be used (Fig. 61).

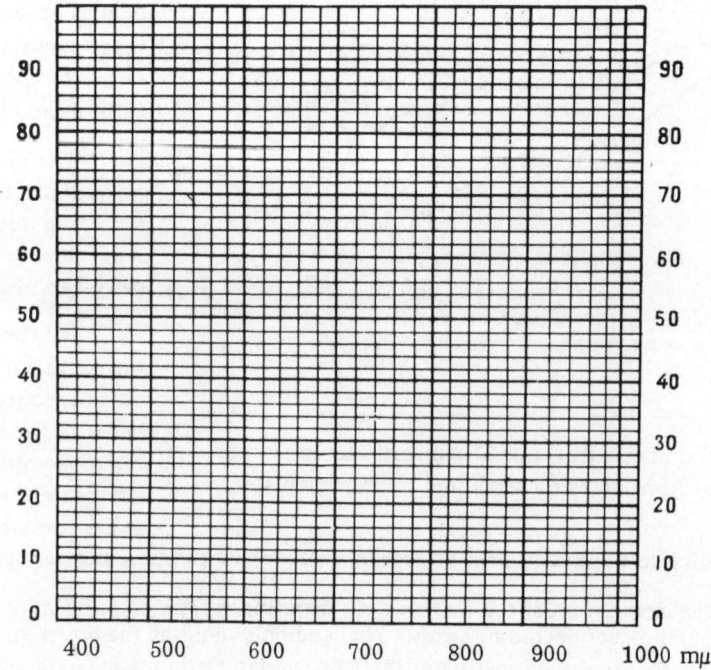

Fig. 61. Graticule for securing equality of scale in wavelength division of oscillograms of compared standard and investigated object.

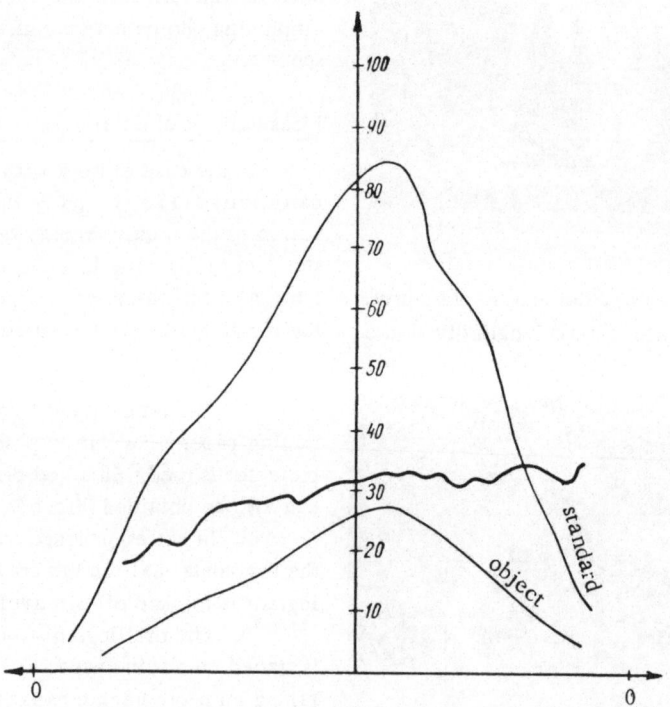

Fig. 62. An example illustrating the plotting of the curve of ρ_λ of an investigated object (center curve). The values of ρ_λ are obtained by dividing the ordinates of the oscillogram of the object by the corresponding ordinates of the oscillogram of the standard.

TABLE 13. Example of Treatment of Oscillograms for the Obtention of the Spectral Luminance Factors of an Investigated Object

Wavelength, mμ	Ordinates of curve, mm		ρ_λ, %
	of standard	of object	
460	14.0	2.5	17.9
480	21.0	4.0	19.0
500	27.0	6.0	22.2
520	33.0	7.0	21.2
540	37.0	9.0	24.3
560	40.5	10.5	25.9
580	43.0	12.0	22.9
600	46.0	13.0	28.3
620	50.0	14.0	28.0
640	55.0	16.0	29.1
660	62.0	17.5	28.2
680	67.0	20.0	29.9
700	71.0	22.0	31.0
720	76.0	24.0	31.6
740	80.0	26.0	32.5
760	83.0	27.0	32.5
780	84.0	28.0	33.3
800	83.0	27.0	32.5
820	76.0	25.0	32.9
840	66.0	23.0	34.8
860	62.5	21.0	33.6
880	59.0	19.0	32.2
900	51.0	17.5	34.3
920	41.0	14.0	35.1
940	34.0	11.5	33.8
960	27.0	8.5	31.5

4. The ordinates for each marked wavelength on the oscillogram of the object are compared with those of the standard and the spectral luminance factors are obtained.

5. The spectral luminance curve of the investigated object is plotted in the system of coordinates (λ, mμ; ρ_λ, %).

Table 13 gives an example of the treatment of an oscillogram obtained in the spectrophotometry of a sand sample from the Repetek region. The nature of the oscillograms for this sand and for the standard and the curve of ρ_λ, are shown in Fig. 62.

It is clear from the descriptions of the processing of spectrograms and oscillograms that these procedures are fairly laborious stages in the obtention of the values of ρ_λ of the investigated objects. In this case the advantage again lies with spectrovisors, since the treatment of oscillograms is simpler (though requiring just as much time) and no additional errors are introduced, as is the case in the use of the methods of photographic photometry.

Of the instruments described above, the one giving the most rapid and accurate measurements of spectral luminance factors is the field photoelectric spectrometer, where the scale is graduated to give ρ_λ as a percentage of the standard. Unfortunately, the field photoelectric spectrometer can only be used in stationary conditions; i.e., this instrument can be used for measurements of individual samples and can serve as a control instrument by means of which the data of spectrophotometry from the air and on the ground can be compared.

Measurements of the spectral luminance factors of geological objects could be carried out more rapidly and more accurately if spectrometers (double-beam) fitted with an improved registering device and capable of excluding errors due to changes in the conditions of illumination during the photometry of the object and standard could be designed.

CHAPTER V

EVALUATION OF SPECTRAL LUMINANCE CURVES AND METHODS
OF GEOLOGICAL INTERPRETATION OF PHOTOMETRIC PARAMETERS

Quantitative Evaluation of Spectral Luminance Curves

As a result of photometry of the investigated rocks we obtain a series of values of spectral luminance factors and from them we plot curves of ρ_λ, which characterize the reflecting properties of the rocks in the measured part of the spectrum. The shape of the obtained curves may be quite different for rocks of different petrographic composition, but characteristic for rocks of similar composition and structure. For instance, the ρ_λ curves of predominantly quartz sands and sands of graywacke composition have an appreciably different shape (see Fig. 16 and Fig. 63), i.e., these rocks can be distinguished from one another by the shape of the ρ_λ curves.

Yet a visual assessment of the shapes of ρ_λ curves is not sufficient and in many cases the confirmation of similarities or differences in ρ_λ curves necessitates their quantitative evaluation. Such an evaluation is also required in connection with the application of certain statistical procedures which can reveal relationships between the photometric parameters of the investigated rocks and their lithological-petrographic composition. Thus, a quantitative evaluation of spectral luminance curves is necessary for the application of photometric data to the solution of various geological problems.

According to requirements, the quantitative evaluation of a spectral luminance curve may be different. For instance, Fig. 64 shows three different ways of evaluating the same ρ_λ curve for a sand sample photometered with the FM-2 universal photometer. The broken line shows the ordinates for ρ_{400} and ρ_{650}, the ratio of which forms the basis of Krinov's classification of the curves of natural objects. The hatched rectangles indicate the mean values of ρ_λ within the color thresholds. The fine line gives the empirical curve plotted from the smoothed values of the observed ρ_λ.

1. As a measure of the ρ_λ curves of natural objects Krinov suggested the coefficient γ, equal to the ratio of the spectral luminance factor in the long-wave region (ρ_{650}) to the spectral luminance factor in the short-wave region (ρ_{400}):

$$\gamma = \frac{\rho_{650}}{\rho_{400}} .$$

On the basis of the values of the coefficient γ Krinov distinguished types of ρ_λ curves characteristic of natural objects (Table 14).

As Fig. 64 shows, the ratio of the two extreme ordinates of the ρ_λ curve only gives the angle between the curve and the wavelength axis. This method of evaluating ρ_λ curves makes it possible to distinguish certain large groups of natural objects, such as water surfaces, where the ratio ρ_{650}/ρ_{400} is less than unity, or the type of ρ_λ curve for vegetable objects, which has a high ratio ρ_{650}/ρ_{400}. This method is characteristically inadequate for the evaluation of the ρ_λ curves of geological objects, since the curves in this case are practically all of the same type.

2. At the start of our investigations, when we tried to correlate the composition of rock samples with their color characteristics, the ρ_λ curves were evaluated by calculations of the mean values of the spectral luminance factors in particular regions of the spectrum. The selected limits of these regions were the color thresholds between the blue and green parts of the spectrum, corresponding to a wavelength of 490 mμ, and between the green and red parts, corresponding to a wavelength of 570 mμ; i.e., all the obtained values of the spectral

68

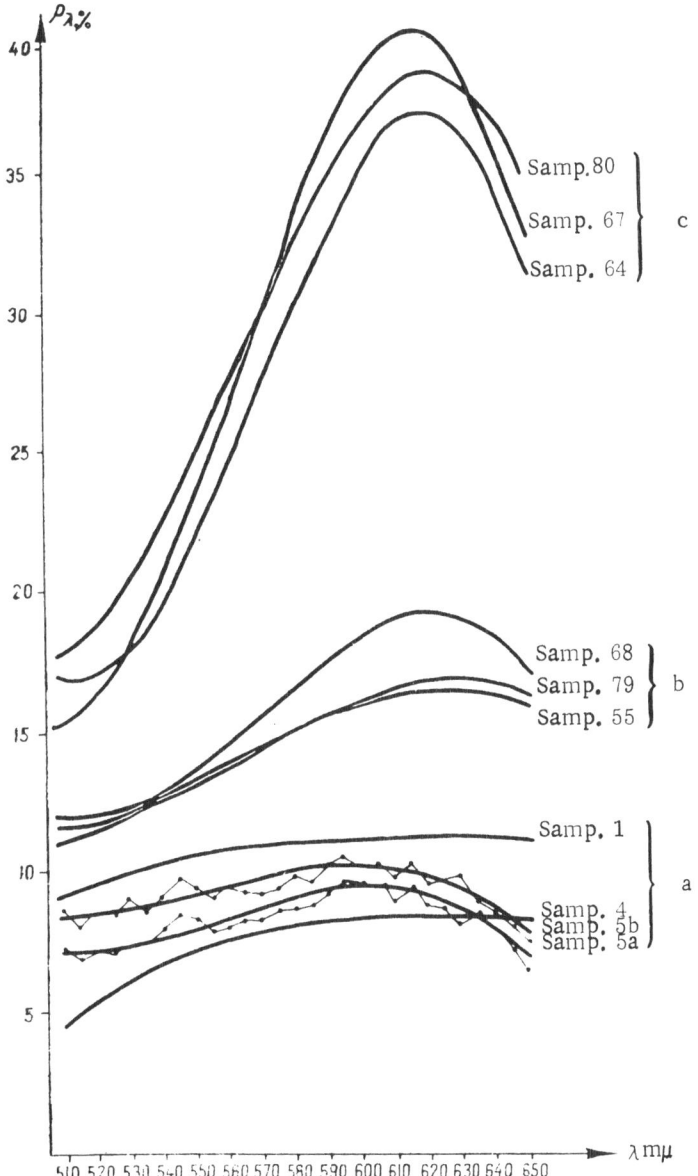

Fig. 63. Types of spectral luminance curves obtained by aerial photometry of sands with different petrographic composition. a) Graywacke; b) polymictic; c) predominantly quartz.

TABLE 14. Types of Spectral Luminance Curves of Natural Objects (according to E. L. Krinov)

Type of ρ_λ curve	Values of ρ_λ, %		γ
	ϱ_{400}	ϱ_{650}	
1	2.4	3.6	1.5
2	7.8	13.5	1.7
3	13.4	34.6	2.6
4	35.7	69.7	2.0

luminance factors of the investigated object were divided into three groups, within which the mean spectral luminance factor was calculated as the arithmetic mean value of ρ_λ in that region of the spectrum. With this method of analysis all the statistical calculations were carried out for the values: $\overline{\rho_1}$, covering the wavelength region 420-490 mμ ; $\overline{\rho_2}$, for the wavelength region 490 to 570 mμ ; and $\overline{\rho_3}$, for the wavelength region 570 to 680 mμ .

The simplicity of the calculations and the clear specific meaning of each $\overline{\rho}$ (mean spectral luminance factor in blue, green, and red regions of spectrum) enabled us to evaluate the effect of a particular constituent of the rock on its spectral luminance, and in some cases to determine the components of the pigment (Romanova, 1958).

The methods of quantitative evaluation of ρ_λ curves are compared in Fig. 64, which shows that the mean ρ_λ within the color thresholds are not sufficiently representative of the properties of the curve. All the information given by the continuous curve is not utilized in this case.

3. To obtain as much information as possible about the properties of the spectral luminance curve from the smallest number of its parameters, Vistelius (1958) suggested using empirical curves, the parameters of which are calculated from the observed ρ_λ.

This method consists in fitting the curves to orthogonal Chebyshev polynomials, for which there are prepared tables, and, hence, the calculation process is considerably simplified.

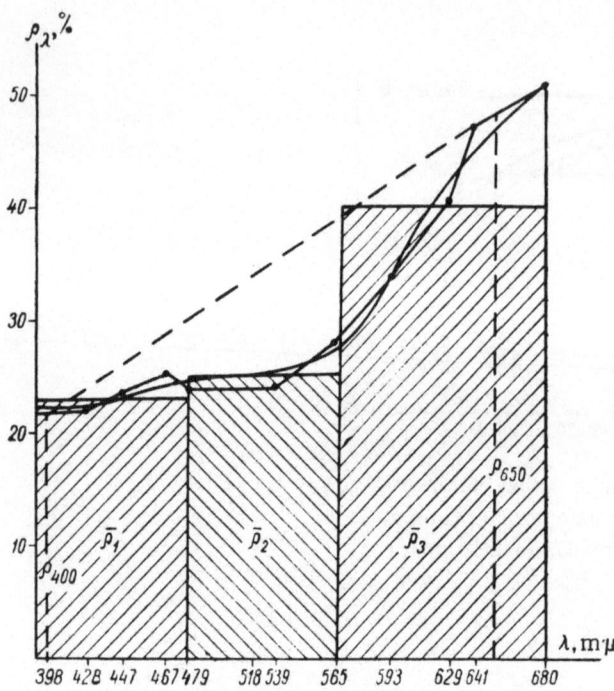

Fig. 64. Evaluation of spectral luminance curves by different methods.

In the conduction of experimental aero-petrographic surveys we usually evaluated the spectral luminance curves by smoothing with Chebyshev polynomials and, hence, we will dwell more fully below on this method of quantitative evaluation.

Quantitative Evaluation of ρ_λ Curves by Smoothing with Chebyshev Polynomials

In the conduction of experimental aero-petrographic surveys we found that the photometric parameters of the investigated deposits could be represented satisfactorily by the values of the polynomial coefficients obtained in the calculation of a third-degree Chebyshev polynomial,

$$\varrho_{(\lambda)} = a_0 + a_1 \lambda + a_2 \lambda^2 + a_3 \lambda^3, \qquad (A)$$

where a_0 is a free term giving the point of origin of the curve on the y axis; a_0 is larger the stronger the reflecting properties of the investigated object; a_1 is the coefficient in the first-degree term of the

TABLE 15. Example of Treatment of a ρ_λ Curve for Division of Region of $\frac{5}{6}$ Observations into 29 Intervals.

ϱ_λ , %	φ_1	φ_2	φ_3	x_i	$f(x_i)$
24.6	−14	+126	−819	1	23.4
24.0	−13	+99	−468	2	24.3
24.4	−12	+74	−182	3	25.2
24.7	−11	+51	+44	4	25.9
26.0	−10	+30	+215	5	26.6
28.4	−9	+11	+336	6	27.1
26.8	−8	−6	+412	7	27.6
28.1	−7	−21	+448	8	28.5
27.8	−6	−34	+449	9	28.4
27.9	−5	−45	+420	10	28.7
28.6	−4	−54	+366	11	28.9
28.3	−3	−61	+292	12	29.2
29.5	−2	−66	+203	13	29.3
29.0	−1	−69	+104	14	29.4
32.6	0	−70	0	15	29.5
30.3	+1	−69	−104	16	29.6
30.8	+2	−66	−203	17	29.6
30.8	+3	−61	−292	18	29.7
29.3	+4	−54	−366	19	29.7
27.3	+5	−45	−420	20	29.8
28.0	+6	−34	−449	21	29.8
29.5	+7	−21	−448	22	29.9
31.1	+8	−6	−412	23	29.9
30.3	+9	+11	−336	24	30.0
30.8	+10	+30	−215	25	30.1
29.6	+11	+51	−44	26	30.3
30.0	+12	+74	+182	27	30.5
29.9	+13	+99	+468	28	30.8
32.7	+14	+126	+819	29	31.0
$\sum_0 = 831,1$	$\sum_1 = 4329$	$\sum_2 = -13229$	$\sum_3 = 49064$		832.8

The obtained sums are divided by the tabulated Chebyshev numbers (App. II) and multiplied by 10 for preservation of the sixth figure.

$K_0 = \frac{1}{2.9} \sum_0$	$K_1 = \frac{1}{2030} \sum_1$	$K_2 = \frac{1}{113274} \sum_2$	$K_3 = \frac{1}{5048784} \sum_3$
286.58621	2.132512	−0.116788	0.009718

$$f(x) = (x - 15) K_1 +$$
$$+ (x^2 - 30x + 150 \cdot 0.83333) K_2 + \qquad (\text{B})$$
$$+ (x^3 - 45x^2 + 540.35x - 1355.25) K_3$$

a_0	a_1	a_2	a_3
286.58621	2.132512	−0.116788	0.009718
−31.98768	3.503640	−0.437310	
−17.527932	5.251121		
−13.170320			
$\sum_{a_0} = 223.90028$	$\sum_{a_1} = 10.887273$	$\sum_{a_2} = -0.554098$	$\sum_{a_3} = 0.009718$

$$f(x) = 223.90028 + 10.887273x - 0.554098x^2 + 0.009718x^3$$

polynomial and gives the slope of the curve relative to the x axis; i.e., in general it determines the steepness of ascent of the ρ_λ curve of the investigated object; a_2 is the coefficient in the square term of the polynomial and indicates the degree of parabolic bending of the curve; a_3 is the coefficient in the cubic term of the polynomial and indicates the curvature of the investigated line, particularly in its right-hand part; with large values of a_3, convex and concave parts of the curve are clearly distinguished.

In addition to providing a measure of the special configurational features of the curves, smoothing of the spectral luminance curves eliminates the effect of random factors affecting its different bends. Such factors may be, for instance, the coarse grain of the emulsion, chance dust particles or scratches on the film of the spectrogram, and so on. For comparison, Fig. 63 shows the rough curves for samples 5[a] and 5[b] plotted from the data of the measured values of ρ_λ and the smooth curves for the same samples plotted from the data obtained by smoothing of the observed values.

Smoothing of curves with Chebyshev polynomials consists in the following operations. All the calculations of the polynomial coefficients must be taken to the sixth figure after the decimal point to ensure an accurate determination of the coefficients of the polynomial (Table 15).

1. The first stage in smoothing of the curves consists in dividing the whole investigated region of wavelengths into a selected number of equal intervals. The values of the spectral luminance factors at the points obtained by division of the whole investigated part of the spectrum are written in the first column.

2. The first column is summed.

3. The products of the numbers of the first column and the corresponding tabulated numbers (Appendix II) in the second column are summed with due regard to sign:

$$\sum_{\varrho} \varphi_1 = \sum_1.$$

4. The products of the numbers in the first column and the corresponding tabulated numbers (Appendix II) in the third column are summed with due regard to sign:

$$\sum_{\varrho} \varphi_2 = \sum_2.$$

5. The products of the numbers in the first column and the corresponding tabulated numbers (Appendix II) in the fourth column are summed with due regard to sign:

$$\sum_{\varrho} \varphi_3 = \sum_3.$$

6. The obtained sums are divided by constants, given in Appendix II, corresponding to the particular number of division points (in this case: 29, 2030, 113274, 5048784) and the corresponding coefficients are found:

$$K_0 = \frac{\sum_0}{29}, \quad K_1 = \frac{\sum_1}{2030}, \quad K_2 = \frac{\sum_2}{113274}, \quad K_3 = \frac{\sum_3}{5048784}.$$

7. The values of the coefficients K_1, K_2, and K_3 are substituted in equation (B) and the parentheses are removed. The coefficient K_0 is not multiplied. It comes in as a component in the sum of the free terms.

8. The coefficients obtained on removal of the three parentheses are the coefficients (a_i) of the smoothing Chebyshev polynomial (A). These coefficients are written in a table.

9. Each column is summed.

10. The equation

$$f(x) = \sum a_0 + \sum a_1 x + \sum a_2 x^2 + \sum a_3 x^3$$

is written down.

---

*It should be noted that when light filters are used, the intervals between them are unequal and, hence, interpolated values of the ρ_λ between adjacent filters must be used in the division of the $\rho\lambda$ curves.

11. The values of this function are calculated by substituting $x = 1, 2, 3, \ldots,$ up to 29. The results are written in the last column [$f(x_1)$, Table 15]. The obtained values should be approximately the same as the numbers in the first column of the same table.

As indicated, we used auxiliary tables of coefficients for calculating the polynomial coefficients of the ρ_λ curves. These tables cover cases of division of the range of observations into 9-31 points at equal intervals (Appendix II).

Evaluation of Photometric Parameters by Plotting Distributions of Observed ρ_λ

The quantitative evaluation of the curves of spectral luminance factors provides us with a set of observations for the photometric parameters of the objects. The observations consist either of the mean ρ_λ within the color thresholds or the polynomial coefficients a_0, a_1, a_2, and a_3.

The solution of problems connected with the aeropetrographic mapping of an investigated region requires a knowledge of the degree of homogeneity (or inhomogeneity) of the values of a particular photometric parameter within the whole set of observations. For this purpose the whole set of observations for each of the considered characteristics is divided into classes and the frequencies in each of the classes are plotted.

On division into classes, which should number about 10-12, the difference between the maximum and minimum value of the considered parameter is divided by the adopted number of classes n:

$$\frac{a_{max} - a_{min}}{n} .$$

The dispersion of the frequencies and the plotting of the distribution curves of the observed value are carried out in the following way (Fig. 65 will serve as an illustration):

1) The values of the considered parameter are plotted on the x axis at the intervals obtained on division into classes.

2) The frequencies of occurrence or the number of observations in each of the classes are plotted on the y axis.

3) From the sums of the frequencies of observations in each of the classes a histogram or frequency polygon is constructed.

If the frequency distribution is symmetrical with one peak (in the center), then it can be assumed, until a deeper analysis is made, that the variations of the values of the considered factor within the investigated area are due to chance.

Practice has shown that the photometric parameters of rocks of homogeneous composition usually give an almost normal (Gaussian) frequency distribution.

On this basis it can be inferred that if the plotting of the frequencies (from a sufficiently large number of observations) gives a distribution with two or more peaks, the variations in the values of the considered parameter are of different origin. Hence, it follows that the photometered rocks in the investigated area have different reflecting properties. In this case we can distinguish

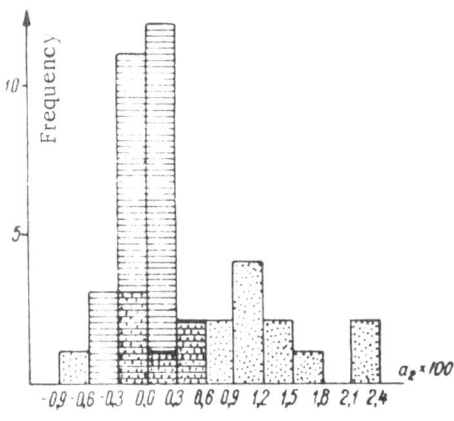

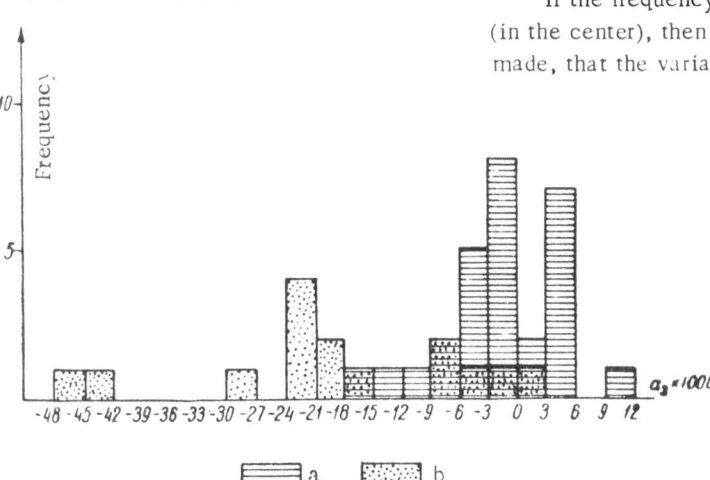

Fig. 65. Distribution of frequencies of a_2 and a_3 for photometered sands from the northwest Caspian region. a) Graywacke sands; b) predominantly quartz sands.

the rocks for which the values of the considered photometric parameter lie within a single one-peak distribution. The boundary of such a distribution can be taken as the value for determining the territorial occurrence of the values of this parameter. As an example Fig. 65 shows the distributions of the values of the polynomial coefficients a_2 and a_3 of the spectral luminance curves for sands from the northwestern Caspian region, obtained by spectrophotometering from the air. From this figure it may be seen that the distributional boundary between values of a_2 and a_3 for graywacke sands and predominantly quartz sands is fairly distinct.

When the homogeneity or inhomogeneity of the distributions of the considered photometric parameter characterizing the spectral luminance of the investigated deposits has been established, the next problem is to discover the geological significance of the measured spectral luminance values.

The existence of a geological meaning of ρ_λ is tested by an investigation of the contingency between the values of the considered photometric parameter and the territorial occurrence of the photometered rock samples, which is required for the demarcation of paleogeographic provinces. The form and degree of relationship are determined by an investigation of the correlation between the values of photometric parameters of the rock and its petrographic-mineralogical composition.

The methods which we used for calculating contingency by χ^2 and linear correlation analysis are adequately expounded in numerous courses of mathematical statistics, among which we can mention those of Smirnov and Dunin-Barkovskii (1959) and van der Waerden (1960). A popular account of the technique of statistical calculations is given by Mitropol'skii (1961).

The method of using these two methods for assessing the relationship between the photometered parameters of rocks and their lithofacies characteristics is described below. Schemes of calculation of contingency criteria χ^2 and correlation coefficients are given.

Method of Calculating Contingency by Means of the χ^2 Criterion

For the practical solution of the problem of revealing an association between the photometric parameters of the investigated rocks, in our case sand deposits, and their lithofacies characteristics, we can use the χ^2 criterion.

The χ^2 test can be used to detect an association between the qualitative attributes of considered objects, as well as between qualitative and quantitative attributes. The behavior of χ^2 does not depend on the type of distribution function of the compared characteristics.

The initial assumption in the application of the contingency test is that the distributions of the observed values are independent. The frequencies of the initial distribution are found from the theorem of multiplication of probabilities.

The test of the initial hypothesis consists in a comparison of the observed frequencies with the frequencies calculated from the initial hypothesis. If the sum of the differences between the calculated and observed

TABLE 16. Scheme for Calculation of Contingency Between Two Attributes by χ^2 Test

Class	y_1	$y_2 \ldots y_n$	Σ
x_1	$\dfrac{[a_{11}]}{(a')}$	$[a_{12}] \ldots [a_{1n}]$	a_{10}
x_2	$[a_{21}]$	$[a_{22}] \ldots [a_{2n}]$	a_{20}
. . . .			
x_m	$[a_{m1}]$	$[a_{m2}] \ldots [a_{mn}]$	a_{m0}
Σ	a_{01}	$a_{02} \ldots a_{0n}$	S

TABLE 17. Example of Calculating Contingency Between Values of Polynomial Coefficient a_3 and Sand Type by the χ^2 Method

| Values of a_3 | Sand type | | $\sum$ |
	Terek	Volga	
< -0.015	1	10	11
> -0.015	30	6	36
$\sum$	31	16	47

$$\chi^2 = \frac{\left(1 - \frac{31 \cdot 11}{47}\right)^2}{\frac{31 \cdot 11}{47}} + \frac{\left(10 - \frac{16 \cdot 11}{47}\right)^2}{\frac{16 \cdot 11}{47}} + \frac{\left(30 - \frac{31 \cdot 36}{47}\right)^2}{\frac{31 \cdot 36}{47}} + \frac{\left(5 - \frac{16 \cdot 36}{47}\right)^2}{\frac{16 \cdot 36}{47}} =$$

$$= \frac{(1 - 7.25)^2}{7.25} + \frac{(10 - 3.74)^2}{3.74} + \frac{(30 - 23.74)^2}{23.74} + \frac{(5 - 12.25)^2}{12.25} =$$

$$= 5.39 + 10.48 + 1.65 + 4.44 = 21.95;$$

$$P\,[(\chi^2) \geqslant 22.4] < 0.001.$$

frequencies falls outside the limits of a particular confidence interval, then the distributions can be regarded as dependent.

The χ^2 criterion is calculated from the formula

$$\chi^2 = \sum \frac{(a - a')^2}{a'},$$

where a are the observed frequencies; a' are the frequencies calculated on the assumption of independence of the distributions.

Having calculated the value of χ^2 we find the probability of $P_k(\chi^2)$, which is the criterion of contingency between the investigated characteristics. A table of confidence limits of the values of χ^2 with f degrees of freedom, taken from Hald (1952), is given in Appendix III. The values are considered to be associated if $P_k(\chi^2)$ < 0.05. The number of degrees of freedom is obtained by multiplying the number of rows minus one by the number of columns minus one.

For the calculation of contingency between two considered attributes, one or both of which may be qualitative, we must construct a frequency distribution table for the investigated attributes and compare the empirical and theoretical frequencies.

We will assume that it is required to calculate contingency between the values of x and y (Table 16). For this purpose we perform the following operations:

1. The considered random values are divided into classes; for instance, in this case the values of x are divided into m classes and the values of y into n classes. In the table we write in the headings of the rows the classes of the first attribute, and in the headings of the columns the classes of the second of the compared attributes. The division into classes is such that each class contains not less than five observations.

2. The whole set of observed values is arranged in the cells of this table so that the two values of the compared attributes correspond to an intersection of a row and a column. We calculate the number of observations (frequencies) in each cell of the table and write it in at the top of the cell. In Table 16 the sums of the observed frequencies are given in square brackets.

3. The frequencies in each row and column are summed. The arrangement of the observed values is checked by summing the final columns and rows and obtaining the same sum S.

4. Under the empirical frequencies in the cells we write in the frequencies obtained on the assumption of independence of the considered attributes (one of these frequencies is given in parentheses in one cell; they have been omitted from the rest). The theoretical frequencies for each cell of the table are obtained from the formula

$$(a_{ij}') = \frac{\sum_i x \cdot \sum_j y}{S} ,$$

where $\sum_i x$ is the sum of the frequencies of the given row; $\sum_i y$ is the sum of the frequencies of the given column; S is the total number of observations. The value of χ^2, which characterizes the contingency between the considered attributes, is equal to the sum of the ratios

$$\chi^2 = \sum \frac{(a_{ij} - a_{ij}')^2}{a_{ij}'} .$$

5. Having obtained the value of χ^2 we verify its confidence limit from Appendix III. If the value of $P(\chi^2)$ is < 0.05 for f degrees of freedom, the hypothesis of contingency between the considered attributes can be regarded as justified; if the value of $P(\chi^2)$ is > 0.05, the hypothesis of independence of the considered attributes is justified. The existence of a relationship can be regarded as confirmed if $P(\chi^2) \leq 0.05$.

An example of calculation of contingency by the χ^2 method for the case of a fourfold table is given in Table 17, where the values of the polynomial coefficient a_3 are compared with types of sand – graywackes from the Terek delta and predominantly quartz sands from the Volga delta.

Investigation of Association Between Spectral Luminance of Rock and Its Petrographic-Mineralogical Composition (Linear Correlation Analysis)

In the case where an association between the photometric parameters of an investigated rock and its lithofacies characteristics has been discovered by the χ^2 test, the nature and strength of the relationship are still unknown.

To investigate the form and strength of the relationship between considered variables we used correlation analysis involving the calculation of paired linear correlation coefficients r. The coefficient of correlation is a numerical measure of the strength and direction of the linear relationship between two random variables x and y.

In our case correlation analysis is required to ascertain the association between the values of the spectral luminance factors of the investigated rocks and their mineralogical composition, i.e., to enable aeropetrographic mapping of rock outcrops from the measured values of spectral luminance factors.

The fundamentals of correlation theory are given in numerous textbooks of mathematical statistics (van der Waerden, 1960, Ch. XIII; Smirnov and Dunin-Barkovskii, 1959, Ch. IX). The application of methods of correlation analysis to the solution of geological problems has been dealt with in a number of papers by Vistelius (1948, 1956) and others.

A description of the procedure for calculating the correlation coefficient in application to the determination of the association between the photometric parameters of rocks and their petrographic-mineralogical composition is given below. Schemes for the calculation of r in the case of a small (less than 30) and large number of observations are given. These procedures can be useful to geologists.

The calculation of correlation coefficients between considered values must begin with a comparison of the correlation tables of the type illustrated in Fig. 66. The position of the observed values of the compared attributes on the coordinate plane (x,y) indicates the presence or absence of an association. For instance, in Fig. 66a the position of the points of the considered attributes on a straight line passing through the origin of coordinates in the plane (x,y) indicates the existence of a direct positive relationship ($r \leq +1$). An equally strong linear relationship, but of a negative nature, is observed if the points lie close to the other diagonal of the correlation table, in which case the value of r is close to -1 (Fig. 66b).

The observed values usually consist of a set of points oriented in some way, and the evaluation of the strength of the relationship in this case requires the calculation of the correlation coefficient in accordance with one of the two schemes (see Tables 18, 20).

When the observed values plotted on the correlation table consist of a nonoriented cluster of points, the absence of a linear relationship is indicated and there is no point in calculating the correlation coefficient in this case (Fig. 66c).

The calculated values of the correlation coefficients are evaluated from the viewpoint of their real significance. The estimation of the significance involves the use of special criteria (see below), for which there are prepared tables.

The scheme for calculation of the linear correlation coefficient in the case of a small number of observations is given in Table 18.

The sequence of operations in this case is as follows:

1. The columns are filled as indicated in Table 18.

2. The sums of the columns in the bottom row are divided by the number of observations n and the means are obtained:

$$\bar{x} = \frac{\sum x}{n}, \qquad \bar{y} = \frac{\sum y}{n},$$

$$\overline{xy} = \frac{\sum xy}{n}, \qquad \overline{x^2} = \frac{\sum x^2}{n}, \qquad \overline{y^2} = \frac{\sum y^2}{n}.$$

a

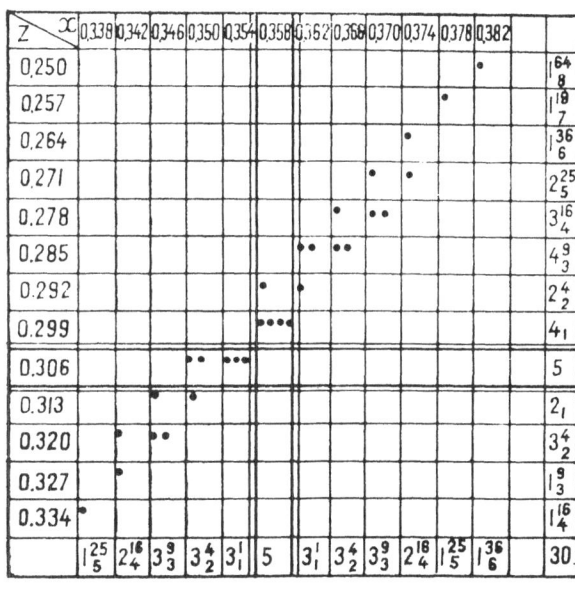

b

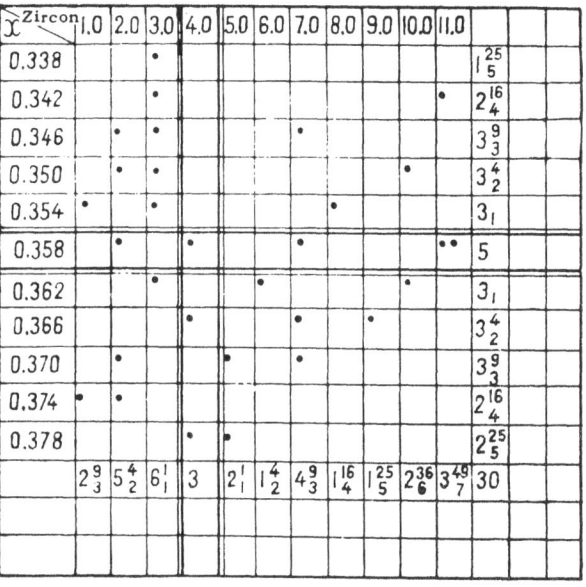

c

Fig. 66. Distribution of observed values of compared attributes. a) Positive relationship; b) negative relationship; c) no relationship.

TABLE 18. Scheme for Calculation of Correlation Coefficient in the Case of a Small Number of Observations

Number of observations	x	y	xy	x^2	y^2	$x+y$	$(x+y)^2$
1	x_1	y_1	$x_1 y_1$	x_1^2	y_1^2	x_1+y_1	$(x_1+y_1)^2$
2	x_2	y_2	$x_2 y_2$	x_2^2	y_2^2	x_2+y_2	$(x_2+y_2)^2$
. . .							
n	x_n	y_n	$x_n y_n$	x_n^2	y_n^2	x_n+y_n	$(x_n+y_n)^2$
	$\sum x$	$\sum y$	$\sum xy$	$\sum x^2$	$\sum y^2$	$\sum(x+y)$	$\sum(x+y)^2$

3. The covariance and standard deviations of x and y are found:

$$\text{Cov}(xy) = \overline{xy} - \overline{x}\,\overline{y},$$

$$S_x = \sqrt{\frac{\sum x^2}{n-1} - \frac{\left[\sum x\right]^2}{n(n-1)}}, \quad S_y = \sqrt{\frac{\sum y^2}{n-1} - \frac{\left[\sum y\right]^2}{n(n-1)}}.$$

4. The correlation coefficient r is calculated:

$$r = \frac{\text{Cov}(xy)}{S_x S_y}.$$

The criterion of the significance of the correlation coefficient will be, depending on the strictness of the requirements,

$$t = Z(r)\sqrt{n-3} \geqslant 2 \quad \text{or} \quad \geqslant 3.$$

In our case the correlation coefficients are calculated in pairs between the value characterizing the reflectance of the rock and the content of the particular component in the rock. As the photometric parameters we can use either $\overline{\rho_1}, \overline{\rho_2}, \overline{\rho_3}$ or any of the polynomial coefficients a_0, a_1, a_2, a_3.

An example of a calculation of the correlation coefficient r between the clay content of a rock and the value of $\overline{\rho_3}$ is shown in Table 19 (the samples analyzed were obtained from the upper part of the red beds on Cheleken Peninsula). This example reveals a direct linear relationship between $\overline{\rho_3}$ and the clay content of the rock. This scheme for calculation of the linear correlation coefficient is used in the case of small number of observations ($n \leq 30$).

A different scheme of calculation, illustrated in Table 20, is used in the case of a large number of observations. The steps in the calculation of correlation coefficients according to this scheme (see Table 20) are as follows:

1. The frequencies are set out in accordance with the division of each of the compared attributes into classes. The classes of the compared attributes are obtained by dividing the difference between the extreme values of the considered attribute into 10 or 12 intervals.

2. Coordinate axes with their center approximately coinciding with the centroid of the tabled points are drawn. The center is taken as the zero of the coordinate system (in Table 20 the new coordinates are distinguished by a double line). The signs of the obtained fields are marked – positive with respect to the diagonal of the origin of coordinates (in Table 20 the signs of the fields are indicated in boxes).

TABLE 19. Example of Calculation of Correlation Coefficient Between Mean Value of Spectral Luminance Factor in Long-Wave Region of Spectrum (x) and Clay Content of Rock (y). (Samples from red beds on Cheleken Peninsula)

x	$\underset{\%\ \times\ 10}{y}$	xy	x^2	y^2	$x+y$	$(x+y)^2$
26	48	1248	676	2304	74	5476
35	656	22960	1225	430336	691	477481
30	312	9360	900	97344	342	116964
33	327	10791	1089	106929	360	129600
30	340	10200	900	115600	570	136900
26	28	728	676	784	54	2916
27	56	1512	729	3136	83	6889
25	58	1450	625	3364	83	6889
27	133	3591	729	17689	160	25600
26	42	1092	676	1764	68	4624
27	245	6615	729	60025	272	73984
25	93	2325	625	8649	118	13924
24	20	480	576	400	44	1936
24	68	1632	576	4624	92	8464
28	117	3276	784	13689	145	21025
25	108	2700	625	11664	133	17689
26	71	1846	676	5041	97	9409
24	46	1104	576	2116	70	4900
30	389	11670	900	151321	419	175561
27	91	2457	729	8281	118	13924
34	637	21658	1156	405769	671	450241
25	21	525	625	441	46	2116
24	37	888	576	1369	61	3721
27	39	1053	729	1521	66	4356
25	59	1475	625	3481	84	7056
26	39	1014	676	1521	65	4225
26	68	1768	676	4624	94	8836
24	26	624	576	676	50	2500
26	15	390	676	225	41	1681
25	18	450	625	324	43	1849

$\sum$ 807	4207	126310	21961	1465011		1740736
Means $\cdots$ 26.9	140.2	4210.33	723.06	48833.7		

$$\sum xy = 126310,\ \overline{xy} = 4210.33,$$

$$S_x = 2.907,\ S_y = 170.79.$$

$$\mathrm{Cov}\,(xy) = 4210.33 - (26.9 \times 140.2) = 438.06;$$

$$r = \frac{438.06}{2.91 \cdot 170.8} = 0.88,\ t = 7.1.$$

$$\sum x^2 + \sum y^2 + 2\sum xy = \sum (x+y)^2,$$

$$\varrho_{y/x} = 51.82.$$

The regression equation is
$$\overline{y}_i = 51.8\,x_i - 1254.$$

3. The first product moment ($\nu_{q,\overline{p}}$) is calculated; this is done by multiplying the frequencies in each cell by the product of the coordinates of the cell. The obtained products are summed with due regard to the signs of the fields and divided by the number of observations.

4. The first moments are calculated:

a) for the rows; the sums of each row are multiplied by their coordinates from the columns. The obtained values are summed and divided by the number of observations $\nu_{\overline{p}}$ in Table 20);

b) for the columns; the sums in each column are multiplied by their coordinates from the rows. The obtained values are summed and divided by the number of observations ($\nu_{\dot{q}}$ in Table 20).

5. The second moments or the dispersions of the compared components are calculated. This is done by:

a) multiplying the sums of the frequencies in each row by the squares of the coordinates from the columns, summing and dividing by the number of observations ($v\frac{n}{\rho}$ in Table 20).

b) multiplying the sums of the frequencies in each column by the squares of the coordinates from the rows, summing and dividing by the number of observations (v''_q in Table 20).

The calculated second moments are equal to

$$\mu_{\bar{\rho}} = v''_{\bar{\rho}} - (v'_{\bar{\rho}})^2, \quad \mu_q = v''_q - (v'_q).$$

6. The first mixed central moment (covariance) is calculated by subtracting the product of corrections to it (v'_q and $v\frac{n}{\rho}$) from the first central moment:

$$\mu_q = v_{q,\bar{\rho}} - (v'_q\, v'_{\bar{\rho}}).$$

7. The standard deviations are calculated; i.e., the square root of the dispersion of each considered component is extracted:

TABLE 20. Example of Calculation of Correlation Coefficient Between Two Variables by the Procedure for a Large Number of Observations

$\bar{\rho}_{570-680}$	Quartz content of sand samples,%											
	>14	>17	>20	>23	>26	>29	>32	>35	>38	>41	>44	
19.4				1_4								1^{16}
21.0			3_6		1							4^9
22.6		$\boxed{+26}$			1	2_2	1_4		$\boxed{-11}$			4^4
24.2		1_3		1_1	1			1_3				4^1
25.8				1	4	1		1				7
27.4	1_3				1	1_1		1_3				4^1
29.0						1_2	2_4	1_6				4^4
30.6					1	1_3		2_9				4^9
32.2									2_{14}			2^{16}
33.8		$\boxed{-4}$						$\boxed{+134}$		1_{25}		1^{25}
>35.4											1_{34}	1^{36}
	1^{16}_4	1^9_3	3^4_2	3^1_1	9	6^5_1	3^4_2	6^4_3	2^{16}_4	1^{25}_5	1^{36}_6	36

$$v_{q,\bar{\rho}} = \frac{26+134-4-11}{36} = \frac{+145}{36} = 4.03; \qquad v'_{\bar{\rho}} = \frac{+43-18}{32} = \frac{15}{36} = 0.42;$$

$$v'_q = \frac{+49-16}{36} = \frac{33}{36} = 0.92; \qquad v''_{\bar{\rho}} = \frac{149+72}{36} = 6.14;$$

$$v''_q = \frac{+165+40}{36} = 5.69; \qquad \mu_{\bar{\rho}} = 6.14 - (0.42)^2 = 5.96;$$

$$\mu_q = 5.69 - (0.92)^2 = 4.84; \qquad S_{\bar{\rho}} = \sqrt{5.96} = 2.44.$$

$$S_q = \sqrt{4.84} = 2.20. \qquad \mu = 4.03 - [0.92 \cdot 0.42] = 4.03 - 0.39 = 3.64;$$

$$r = \frac{3.24}{2.20 \cdot 2.44} = \frac{3.64}{5.37} = +0.68; \quad t = Z\sqrt{n-3} > 3.$$

$$S_{\bar{\varrho}} = V\overline{\mu_{\bar{\varrho}}}, \; S_{\bar{q}} = V\overline{\mu_q}.$$

8. The correlation coefficient is calculated. This is done by dividing the value of the first central mixed moment by the product of the standard deviations of the compared components (variables):

$$r = \frac{\mu}{S_1 S_2}.$$

9. The significance of the obtained correlation coefficient is estimated. The significance criterion t in the investigation of relatively symmetrical single-peak distributions is the inequality

$$t = Z(r) \cdot V\overline{n-3} \geqslant 2 \quad \text{or} \quad \geqslant 3,$$

(depending on the width of the confidence interval)

where Z is Fisher's transformation, for which there are special tables (Appendix IV).

When the series of linear correlation coefficients between the compared components or properties of the rock has been calculated we can estimate the strength of the relationship between them, i.e., what change in the content of one component in the analyzed rock affects the mean content of the other component or property of the rock. To cover the whole complex of relationships between the considered components and to illustrate all the sets of associations revealed, we can compile a summary matrix of correlation coefficients (see Table 36), from which it is easy to determine mineral associations or the nature of the relationship between the mineral components of a rock and its spectral luminance factors. If we have a sufficiently large number of observations and significant relationships between the spectral luminance of a rock and its petrographic-mineralogical composition, we can solve the reverse problem; i.e., by evaluating the spectral luminance curves of investigated rocks lying within a mapped area, we can determine the lithofacies type of the photometered rock.

The problem could be treated more thoroughly and accurately by the use of multiple correlation coefficients, which would enable us to determine the combined effect of all the components on the ρ_λ of a rock. It would be possible to determine the effect of one component of the rock on the photometric parameter associated with it by using regression equations. In the case of a strong linear relationship we could use the value of the photometric parameter to determine the mean content of a particular component significantly associated with the photometric parameter.

The procedure for calculation of the regression equation, according to Hald (1956), is exemplified at the end of Table 19 by the calculation of $\bar{\rho}_3$ for rock with known clay content.

All the above methods of revealing association can be used for the solution of specific problems. The technique of calculating correlation coefficients and regression coefficients is adequately and lucidly expounded in Mitropol'skii's book (1961).

Thus for the determination of association between the photometric parameters of rock and their lithofacies characteristics we used the χ^2 method, which reveals association between quantitative and qualitative attributes, and correlation analysis, which provides an estimate of the linear statistical relationship between the considered components.

CHAPTER VI

CORRELATION OF SPECTRAL LUMINANCE WITH LITHOFACIES
CHARACTERISTIC OF DEPOSITS

Before going on to the account of the results of aerial spectrophotometric rock surveying, which constitutes the basis of aeropetrographic mapping, we must discuss the results of laboratory investigations on association between the photometric parameters of sandy-silt deposits and their lithofacies characteristics, which are considered in this chapter.

By the lithofacies characteristics of the deposits we mean the whole complex of properties of deposits formed in particular conditions of sedimentation.

To determine the nature of the association between the reflecting properties of sandy-silt deposits and their lithofacies characteristics, we photometered a large number of samples collected from terrigenous deposits of various age in different regions of West Turkmenia. The aim of these investigations was to demonstrate the possibility of mapping facies from the measured values of ρ_λ of different representative samples of a particular complex of deposits.

We investigated three stratigraphic complexes in different regions of West Turkmenia: Middle Jurassic deposits, terrigenous Cretaceous deposits, and Tertiary red beds (Fig. 67). In each of these complexes we

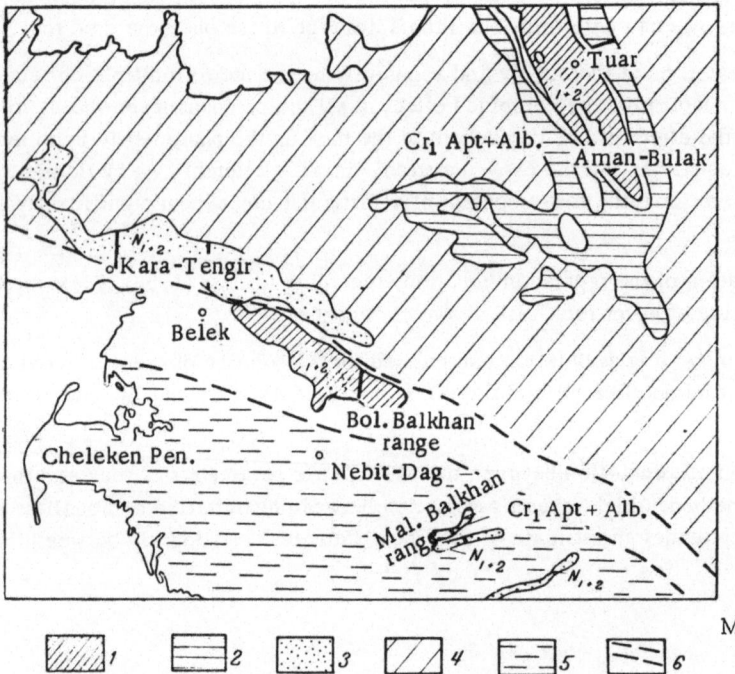

Fig. 67. Position of investigated sections (heavy lines) within the structural regions of West Turkmenia. 1) Middle Jurassic deposits; 2) terrigenous Cretaceous; 3) Neogene deposits; 4) region with platform type of sedimentation; 5) region with geosynclinal type of sedimentation; 6) transitional region between 4 and 5. The broken lines indicate the hypothetical boundary (according to Godin) between the epiHercynian platform and the Alpine geosyncline.

investigated two parallel synchronous sections, one from a region where the deposits had accumulated in approximately geosynclinal conditions, and the other from a region where the deposits had accumulated in conditions close to typical platform conditions.

The problem was regarded as solved if externally similar deposits of approximately the same petrographic composition from different regions had different spectral luminance factors and if the differences in the parameters of the $\rho\lambda$ curves in the χ^2 test were significant.

The methods of investigating the compared sections were the same in all cases and were as follows:

1. In the region marked off for photometric measurements we selected an area which was best exposed and compiled the usual stratigraphic description of the section. We took samples of silty or sandy beds at approximately equal intervals of thickness. If such beds were absent we used the sand fraction from conglomerate beds.

The weight of each sample was 2-2.5 kg. The sample was quartered and from the remaining portion 100-120 g was taken for the photometric measurements, 10-12 g for the preparation of microsections, 20 g for chemical analysis, and 50-100 g for granulometric analysis. The rest of the sample was used to obtain a gray heavy residue by triple washing in a pan or dish. All these operations were performed in the field.

2. In the laboratory the samples were subjected to various analyses and measurements.

A. The photometric measurements of the samples were performed on the FM-2 universal photometer. From the results of the measurements we plotted the curves of $\rho\lambda$ and then smoothed these curves by the above-described method to obtain the polynomial coefficients.

B. The granulometric analyses were carried out by a simplified Williams method. A 50- or 100-g batch was flooded with 5% HCl and heated. More of the same acid was then added until effervescence completely ceased. The sample left on the filter was washed with distilled water and dried in a desiccator. The difference in weight, expressed as a percentage, was taken as the carbonate content of the sample. The dry residue of the sample was ground up in a porcelain dish with a rubber-tipped pestle and then transferred to a liter jar. The clay fraction was determined as the loss in weight after elutriation of the sample by decantation of the upper layer of the suspension after 5 min standing, until the water above the sample was completely clear. The residual sandy-silt fraction was dried and passed through sieves. The sieved fractions were weighed on an analytical balance and the calculated percentage proportions of the fractions were set down in the form of tables of the granulometric composition of the rocks.

C. The mineralogical analysis of the samples consisted of an investigation of the composition of the samples in thin sections with a count of the fragments of quartz, plagioclases, predominantly potassic feldspars, micas, and fragments of magmatic and sedimentary rocks. The minerals of the heavy fraction were investigated under a binocular microscope and in immersion mounts. The heavy residue was separated into fractions in bromoform. The heavy fraction was then separated with an ordinary horseshoe magnet to obtain the magnetic fraction and with a BIT-2 electromagnet to extract the electromagnetic fraction. The quantitative determination of the magnetic, electromagnetic, and nonmagnetic fractions in the heavy fraction was performed separately in sample troughs. The mineralogical composition of the heavy fraction was expressed in nominal percentages relative to the weight of the fraction but without corrections for the densities of the individual heavy minerals. The results of the quantitative mineral analysis of the investigated rocks are given in the form of separate tables for each section.

D. The chemical analyses of the samples were selective and were performed only for those components which would be likely to affect the photometric parameters of the investigated rocks. A quartered batch of 1-3 g was boiled in a 10% HCl solution for 1 hr. In the obtained acid extract the total soluble iron was determined by reduction of all the soluble iron to ferrous iron with hydrogen in a Jones reducer with zinc. The completeness of the reduction of iron was checked with ammonium thiocyanate $NH_4(CNS)$. After the determination of the total reducible iron the total iron content was converted to Fe_2O_3. Ferrous iron FeO was also determined from the hydrochloric acid extract by titration with potassium bichromate ($K_2Cr_2O_7$) in the presence of diphenylamine as an indicator. The content of soluble ferric iron Fe_2O_3 was calculated as the difference between the total iron content, converted to Fe_2O_3, and the ferrous iron content, also converted to Fe_2O_3. All the calculations were performed on a dry batch with due regard to hygroscopy. The results of the analyses were set down in tables for each of the compared sections.

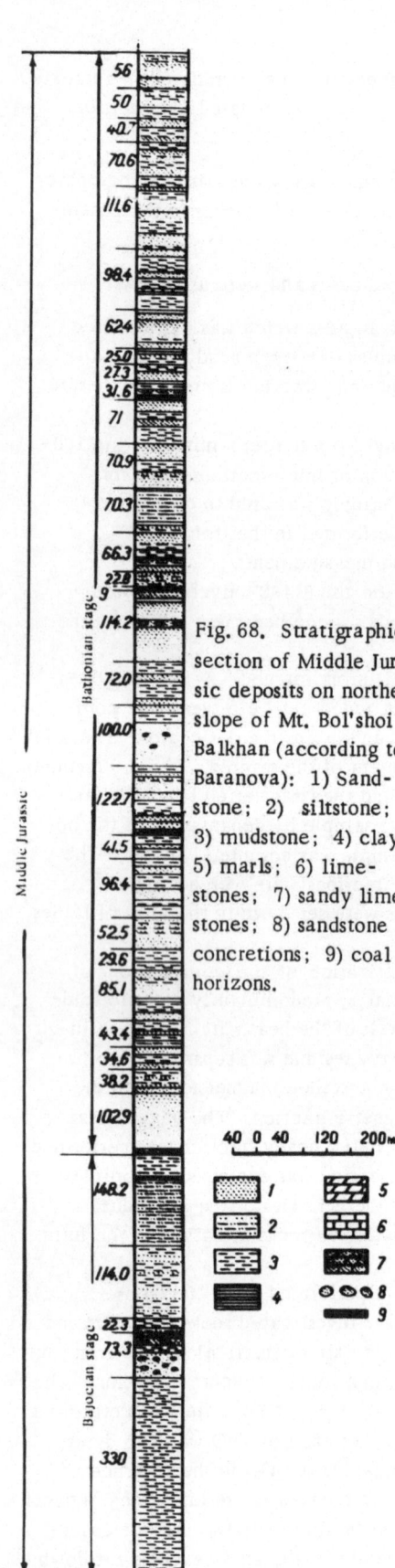

Fig. 68. Stratigraphic section of Middle Jurassic deposits on northern slope of Mt. Bol'shoi · Balkhan (according to Baranova): 1) Sandstone; 2) siltstone; 3) mudstone; 4) clays; 5) marls; 6) limestones; 7) sandy limestones; 8) sandstone concretions; 9) coal horizons.

E. The data of all the analyses of the investigated rocks (granulometric, mineralogical, chemical) were compared with the photometric data for these samples. The values of the spectral luminance factors were treated by the method described in Chapter V.

Below we give a description of the three stratigraphic comlexes of West Turkmenia – Middle Jurassic, terrigenous Cretaceous, and Neogene deposits – the samples from which showed the geological meaning of the measured spectral luminance factors and, hence, the possibility of facies mapping of terrigenous deposits from their reflecting properties.

Descriptions of each stratigraphic complex are given from two compared sections representing the geosynclinal and platform types of sedimentation. In the description of the sections we used the works on the stratigraphy of Jurassic and Cretaceous deposits of West Turkmenia (Baranova, 1956; Kravets, 1954) and also on the spectral luminance of terrigenous Cretaceous deposits (Vistelius, 1958).

Middle Jurassic Deposits of West Turkmenia

Photometric investigations of the Middle Jurassic deposits in West Turkmenia were made in two sections – one on the northern slope of Mt. Bol'shoi Balkhan, where the deposits approximate the geosynclinal type, and the other in the Tuarkyr region, where they are represented by platform facies. The samples for photometry were all taken from strata of sandy-silt composition, so that the whole investigated section was more or less uniformly represented. The investigated sections and rocks are described fairly fully so that the reader can understand the special features of the structure of the section and the petrography of the photometered rocks.

Middle Jurassic Deposits of Bol'shoi Balkhan *

The section is built up from bedrocks exposed on the sides of the valley of a temporary stream flowing from the head of a ravine near the summit of Kosha-Dzhul'ba toward the wells of Kara-Iman; the section runs almost north-south. Figure 68 shows the stratigraphic column of the section, the description of which follows (from bottom to top):

1. Compact black mudstones with bands of greenish-gray sandstones. The set is uniform, thick, and contains a fauna of Oxytoma sp., Lima sp., and Macrodon sp. Thickness 330 m.

2. Unit of black mudstones with bands of conglomerate-like sandstones containing marl and siderite concretions with a fauna of Phylloceras cf. heterophylloides Opp., Macrodon sp., Nucula sp., and Lucina sp. Thickness 73.3 m.

3. Unit of dark greenish-gray compact siltstones alternating with darker mudstones. Thickness 22.3 m.

4. Black mudstones with bands of siltstones and sandstones. The mudstones contain concretions with a fauna of Nucula sp. Thickness 114 m.

---

* The description is taken from Baranova (1956).

5. Black mudstones with intercalations of siltstones and rare beds of calcareous sandstones. The upper part of the bed of clayey siltstones contains fragments of belemnites and Perisphinctes tenuissimus Siem. The top of the unit is composed of mudstones alternating with gray fine-grained platy sandstones. Thickness 148 m.

The boundary of the Bajocian runs along the foot of the first thick bed of sandstones (at least 1 m thick). The total thickness of the predominantly argillaceous strata assigned to the Bajocian stage is 675 m. According to N.P. Luppov's data, the thickness of deposits of Bajocian age in the region of the wells of Kara-Chagyl reaches 1000 m.

Above these strata lies the gray mudstone-sandstone series assigned to the Bathonian stage.

6. Fine-grained sandstones with clayey-calcareous cement, gritstones with white quartz and black siliceous pebbles. Mudstones with remains of plant detritus. Thickness of set 102.9 m.

7. Mudstones alternating with massive and finely stratified silty sandstones. Thickness 38.2 m.

8. Pale yellowish gray silty massive sandstones with water-ripple marks and with cross bedding. The cement of the sandstones is chloritic-clayey or carbonate. The unit contains a fauna of Tancredia sp. and Pleuromya sp. Thickness 34.6 m.

9. Greenish gray silty mudstones alternating with cross-bedded pale sandstones. The cement of the sandstones is clayey in the lower layers of the unit and carbonate in the upper. Thickness 43.4 m.

10. Yellowish gray, weakly cemented fine-grained sandstones with two layers of silty mudstone. Sandstones cemented with carbonate cement at the bottom, with calcareous-clayey cement in the center, and with choritic-siliceous-clayey cement at the top. The unit contains an abundant fauna of ammonites and trigonias Pseudomonotis ex gr. echinata Smith, Posidonia sp., Trigonia sp. ind., Goniomya baysunensis Boriss. Thickness 85.1 m.

11. Unit consisting of an alternation of tobacco-colored mudstones, gray silty sandstones, massive sandstones and clayey platy sandstones. Pelecypods in a poor state of preservation are found. Thickness 29.6 m.

12. Fine-grained sandstones with bands of clayey sandstones and mudstones. The last contain a fauna of Oppelia cf. fusca Quenst. and Macrodon cf. babkhanensis Peel. Thickness 52.5 m.

13. Alternating beds of dark gray mudstones, clayey lumpy siltstones, and fine-grained sandstones. Bands of coarse-grained sandstones up to 20 cm thick and containing a fauna of Pseudomonotis ex. gr. echinata Smith., Pseudomonotis sp., and Pecten sp. are found. Thickness 96.4 m.

14. Pale massive sandstones alternating with dark gray mudstones containing a fauna of Belemnopsis sp. and Nucula sp. Thickness 41.5 m.

15. Massive yellowish-gray fine-grained sandstones alternating with mudstones. The unit contains an abundant fauna of Pseudomonotis cf. echinata Smith., Macrodon sp., Nucula sp., and Trigonia cf. signata Ag. Thickness 122.7 m.

16. Greenish-gray mudstones with thin bands of platy sandstones and layers of massive fine-grained sandstones. A fauna of ammonites and gastropods. Thickness 100 m.

17. Dark-gray greenish mudstones with two layers of pale massive cross-bedded fine-grained sandstones. Thickness 72 m.

18. Medium-grained yellowish-green massive sandstones with lenses of ferruginous coarse-grained sandstones containing trigonias. Thickness 114.2 m.

19. Coal-bearing horizon with clays containing a fauna of Modiola sp. and Myopholas cf. ocumonate Sow. Thickness 9 m.

20. Gray mudstones alternating with platy greenish-gray silty sandstone with a fauna of Nucula sp. and Astarte sp. Thickness 22.8 m.

21. Yellowish-green massive sandstones. Mudstones with a fauna of Leda sp. and Nucula sp. Thickness 66.3 m.

22. Greenish-gray mudstones. Fine-grained sandstones with casts of pelecypods Goniomya sp. Thickness 70.3 m.

23. Fine-grained greenish-gray sandstones. Massive mudstones with vertical worm trails. Thickness 70.9 m.

24. Unit of dark gray mudstones alternating with yellowish-gray sandstones containing a fauna of Nucula sp. Thickness 71.7 m.

25. Second coal-bearing horizon and unit of greenish-gray mudstones. Thickness 31.6 m.

26. Fine-grained massive greenish-gray sandstones with bands of platy sandstones. Thickness 25.0 m.

27. Greenish-gray mudstones, clays, carbonaceous clays with remains of flora. Thickness 25.0 m.

28. Homogeneous massive clayey siltstones, mudstones, massive sandstones with a fauna of Entolium cf. ivanovi Pčel., Opis sp., Lucina cf. bellona Orb. Thickness 62.4 m.

29. Silty greenish-gray mudstones with bands of platy limestones, massive fine-grained sandstones with worm trails. Thickness 98.4 m.

30. Pale gray clays alternating with massive cross-bedded sandstones (beds up to 19 m thick). Lenses of coarse-grained sandstones containing a fauna of Entolium sp. and Trigonia cf. formos Lyc. are found within the sandstones. Thickness 111.6 m.

31. Unit consisting of greenish-gray mudstones alternating with cross-bedded sandstones. The fauna found in these layers was identified by L.V. Sibiryakova as Pseudomonotis sp. and Entolium cf. ivanovi Pčel. Thickness 70.6 m.

32. Massive yellowish-gray fine-grained sandstones, becoming dirty green and with a greater amount of cement higher in the section. Thickness 40.7 m.

33. Unit of mudstones alternating with sandstones containing a fauna of Pseudomonotis sp. and Modiola sp. Thickness 50 m.

34. Massive pale-gray fine-grained sandstones, quartz grains well sorted, cross bedding apparent. Abundant fauna of pelecypods: Lucina sp., Pleuromya sp., Goniomya sp., and Goniomya baysunensis Boriss. Thickness 56 m.

The overlying deposits are greenish-gray massive calcareous sandstones with a fauna of Hecitcoceras cf. salvadorii (Bor.) and are assigned to the Callovian stage.

The total thickness of the deposits of the Bathonian stage in the described section is 1818 m. The total thickness of the Middle Jurassic deposits in the investigated section is 2475 m.

Thus the investigated section of Middle Jurassic deposits exposed on the northern slope of Mt. Bol'shoi Balkhan consists of thick sets of rocks of terrigenous origin. The lower part of the section is composed mainly of mudstones and belongs to the Bajocian stage, while the upper, predominantly sandstone part belongs to the Bathonian stage. The boundary between them runs along the bottom of the first thick layer of sandstone. The first Parkinsonia horizon, situated in the top half of the upper Bajocian and with a thickness of 300 m, can be

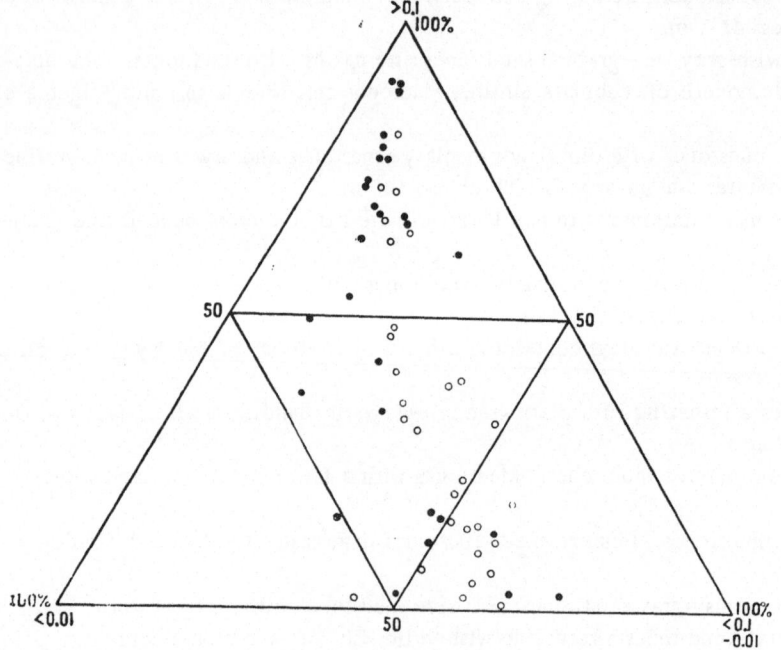

Fig. 69. Granulometric composition of samples of Middle Jurassic deposits in West Turkmenia. Open circles refer to the Bol'shoi Balkhan section; filled circles refer to Aman-Bulak section.

distinguished. Among the beds with the ammonites Parkinsonia cf. doneziana Boriss. and Phylloceras zignadianum Orb., the second Parkinsonia horizon, assigned to the Bathonian, can be distinguished. The great thickness and uniformity of the deposits of terrigenous origin are characteristic features of deposits of the Central Jurassic section exposed on the northern slope of Mt. Bol'shoi Balkhan.

For photometric measurements we took samples from beds of sandstones and siltstones so that the described section was more or less uniformly represented. Twenty-nine samples were taken from the section.

The granulometric composition of the samples taken from the Bol'shoi Balkhan section is given in Table 21 and illustrated in the form of points on a classification triangle (Fig. 69).

TABLE 21. Granulometric Composition and Carbonate Content of Samples from Middle Jurassic Deposits in Bol'shoi Balkhan Section (West Turkmenia)

No. of sample	Carbonate content	Fractions, mm										Total
		>0.59	0.59–0.42	0.42–0.29	0.29–0.21	0.21–0.149	0.149–0.105	0.105–0.074	0.074–0.053	0.053–0.01	<0.01	
1	10.3	—	—	—	0.1	4.5	26.3	11.6	10.9	14.9	31.3	99.6
2	9.2	—	—	—	—	0.6	21.6	18.9	15.0	14.7	29.2	100.0
3	6.7	—	—	—	0.1	0.3	14.1	16.0	19.3	16.3	33.7	99.8
4	9.7	—	—	—	—	0.8	31.2	22.6	15.2	10.6	19.2	99.6
5	11.7	—	—	—	—	0.4	9.1	10.9	22.3	24.0	32.7	99.4
6	18.4	—	—	—	0.1	0.2	1.3	6.7	25.1	27.1	38.7	99.2
7	13.6	—	—	—	0.1	0.1	5.0	14.3	24.2	23.2	32.5	99.4
8	10.0	—	—	—	—	0.3	13.5	15.9	7.6	30.3	32.0	99.6
9	7.6	—	—	—	0.0	3.6	35.1	14.6	13.5	10.1	23.2	100.1
10	16.6	—	—	—	—	0.3	14.3	17.6	22.4	18.0	27.0	99.6
11	6.4	—	—	—	—	1.0	17.4	15.0	15.4	19.4	31.5	99.7
12	3.4	—	—	—	—	0.1	0.7	3.0	12.7	28.0	55.5	100.0
13	8.4	—	—	—	0.0	0.1	0.6	14.4	23.7	27.3	33.6	99.8
14	7.6	—	—	—	0.0	3.3	29.2	11.8	10.3	12.8	32.4	99.8
15	14.3	—	—	—	—	0.1	14.0	14.8	23.4	16.8	30.6	99.7
16	7.9	—	—	—	0.1	0.5	3.5	8.2	27.0	24.3	36.3	99.9
17	8.6	—	—	—	—	0.2	5.9	16.6	21.9	12.5	42.7	99.8
18	25.0	0.4	0.4	—	10.9	45.8	12.1	3.6	4.1	3.1	16.8	100.0
19	7.7	—	—	—	0.1	12.3	33.7	8.5	9.6	8.0	27.6	99.9
20	5.2	—	—	—	14.4	36.8	13.3	3.7	6.1	4.8	19.3	99.7
21	6.0	—	—	—	0.4	8.7	30.4	17.0	13.8	8.8	20.8	99.9
22	8.0	—	—	—	0.1	1.6	9.8	19.3	25.6	13.5	29.7	99.6
23	8.8	0.1	0.1	—	2.0	36.9	23.5	5.9	8.5	5.2	17.3	99.9
24	14.4	—	—	—	0.2	8.8	25.6	11.4	12.6	9.5	31.4	99.5
25	24.6	—	—	—	—	9.8	36.7	9.1	9.1	7.5	27.4	99.6
26	1.1	—	—	—	1.8	48.1	20.8	3.5	5.4	4.2	15.6	99.5
28	1.6	—	—	—	0.7	6.8	33.7	8.3	11.3	9.2	30.0	100.1
29	19	—	—	—	0.2	35.2	27.4	5.1	6.1	5.9	20.0	99.9
30	18	0.3	2.1	—	17.8	27.7	9.6	2.8	3.8	3.3	10.0	99.6
$\bar{X}$	10.7					10.17	17.9	11.4	14.7	14.2	28.6	
S	6.05					15.5	11.34	5.78	7.28	8.23	9.6	

The mineralogical composition of the rocks from counts in thin sections is shown in Table 22 and illustrated in the form of points on the classification triangle of Dapples, Krumbein, and Sloss (Fig. 70).

As these data show, the investigated rocks are typical graywackes with silty-sand structure and fine-banded texture. The fragmentary material is slightly rounded and in some cases the grains are oriented along the bedding planes. The cement is interstitial or of the contact type, carbonate or clayey in composition, and makes up 18-20% by volume. The fragmentary material consists mainly of fragments of granophyric and medium-grained granites, many quartz-porphyries and microdiorites, with occasional fragments of keratophyres and lamprophyres. Fragments of sedimentary rocks are rarer and much more rounded, and they are sometimes coated with a thin film of ferric hydroxide.

The rock-forming minerals of the light fraction have the following characteristics:

The great bulk of the quartz occurs in the form of large angular grains, which sometimes contain inclusions of feldspars (quartz from granites), or in the form of grains with mosaic extinction and inclusions of fine scales of mica (veined quartz).

The predominantly potassic feldspars are represented by pale-colored grains of perthitic microclines, and plagioclases of the acid series by relatively fresh brownish-gray grains.

Micas occur in the form of broadly tabular, deformed biotite grains of a brownish or greenish-brown color, which are usually replaced by green chlorite.

TABLE 22. Petrographic Composition of Samples from Middle Jurassic Deposits in Bol'shoi Balkhan Section (West Turkmenia)

No. of sample	Grain size, mm	Content of fragments, % (by volume)				
		feldspars	quartz	mica	rock fragments	cement
1	0.05—0.15	16.3	28.3	4.2	33.6	17.6
2	0.02—0.15	11.5	24.8	5.0	49.5	9.2
3	0.01—0.15	20.8	28.5	6.8	43.9	9.5
4	0.02—0.20	17.2	25.8	6.1	40.7	10.2
5	0.01—0.10	9.3	22.0	2.1	39.6	26.0
6	0.01—0.07	9.8	19.1	3.4	29.4	38.3
7	0.05—1.10	16.9	31.1	4.4	29.3	18.3
8	0.02—1.10	20.3	27.8	4.7	27.8	19.4
9	0.02—0.20	16.4	27.4	6.0	48.0	2.2
10	0.01—0.10	14.1	17.7	2.5	44.3	21.4
11	0.01—0.2	21.6	24.4	6.9	31.0	16.1
12	0.01—0.1	14.9	13.4	10.1	18.7	42.9
13	0.02—0.11	11.4	23.2	4.3	36.4	24.7
14	0.05—0.50	24.1	24.1	6.2	40.1	5.6
15	0.01—0.10	10.5	25.0	3.5	29.9	31.1
16	0.01—0.07	7.7	26.2	3.6	46.3	16.2
17	0.02—0.15	12.7	25.9	2.2	31.0	28.2
18	0.05—0.7	12.4	27.1	1.7	30.0	28.8
19	0.05—0.30	19.3	19.0	4.3	48.1	9.3
20	0.05—0.20	19.5	32.2	2.0	30.8	15.5
21	0.05—0.20	18.5	32.1	4.2	38.5	6.7
22	0.02—0.10	9.5	20.1	5.7	32.3	32.4
23	0.02—0.50	16.3	23.9	5.1	50.2	4.5
24	0.05—0.15	11.4	21.3	3.1	36.0	28.2
25	0.02—0.15	11.1	15.2	2.1	29.3	42.3
26	0.05—0.25	16.3	31.3	2.6	47.8	2.2
27	0.01—1.3	19.4	33.1	2.7	35.4	9.4
29	0.05—0.25	28.0	27.5	6.0	33.7	4.8
30	0.10—0.70	23.0	43.6	2.6	29.4	1.4
$\overline{X}$		15.0	25.5	4.3	36.4	17.6
S		5.0	5.9	1.9	8.2	12.1

TABLE 23. Mineralogical Composition of Heavy Fraction from Middle Jurassic Deposits of West Turkmenia (Bol'shoi Balkhan Section)

Mineral	1	2	3	4	5	6	7	8	9	10	11	12	13	14	15	16	17	18	19	20	21	22	23	24	25	26	27	28	29	30	$\bar{X}$	S
Rock fragments	12.6	11.5	4.3	18.7	23.1	18.9	4.8	9.0	15.6	13.2	52.7	16.3	9.8	8.6	—	8.5	40.6	16.5	30.9	—	5.9	7.6	—	19.0	47.8	25.2	—	22.8	14.0	—	16.3	
Carbonates	0.5	0.3	0.2	—	0.5	—	0.7	R.g.	0.2	—	—	2.2	—	—	0.4	31.3	—	R.g.	R.g.	R.g.	5.04	R.g.	13.1	0.08	0.08	—	—	—	—	0.5	2.0	
Barite	—	—	—	—	—	—	—	—	—	—	—	3.5	—	—	—	—	—	—	—	—	—	—	—	—	—	—	—	—	—	—	0.1	
Apatite	2.5	8.0	3.7	2.9	1.4	1.7	3.3	11.9	2.3	3.8	2.2	0.6	1.2	4.3	R.g.	2.0	2.1	3.2	4.6	—	1.9	2.1	0.6	1.75	2.4	—	—	11.5	0.7	—	3.0	
Monazite	—	—	—	—	—	0.2	—	—	—	—	—	—	—	—	—	0.4	—	—	—	—	—	—	—	—	—	—	—	—	—	—	0.01	
Sphene	—	—	—	—	—	—	—	R.g.	—	—	—	0.1	—	0.2	—	0.5	—	—	—	—	0.2	0.3	—	R.g.	0.98	0.16	—	—	—	R.g.	0.07	
Chlorite	1.9	1.7	0.3	1.1	1.6	0.8	0.7	0.4	1.5	0.3	2.6	0.7	—	—	—	0.8	0.6	0.3	0.85	0.7	0.4	0.7	0.4	0.2	—	—	—	0.8	2.6	—	0.8	
Biotite	—	—	—	—	—	—	—	0.03	—	—	—	—	—	—	R.g.	0.4	—	0.3	—	—	—	—	—	R.g.	—	—	—	—	0.7	—	0.06	
Tourmaline	0.3	R.g.	0.3	0.2	0.4	0.7	0.2	0.7	1.6	0.7	0.3	2.6	1.4	—	1.5	1.5	1.0	R.g.	2.3	1.0	—	1.1	0.4	—	1.8	0.9	—	—	—	—	0.8	
Garnet	9.2	11.7	3.5	2.2	3.6	0.9	5.5	2.7	2.1	3.5	5.9	1.3	4.1	4.0	9.3	0.4	2.5	—	2.0	0.4	1.0	—	—	3.4	—	5.3	—	—	—	—	3.1	3.1
Epidote	6.6	9.8	6.8	3.8	9.4	6.9	11.9	10.0	15.1	27.8	15.9	17.9	7.9	12.2	10.0	2.0	6.8	3.8	10.0	10.2	5.24	26.0	7.4	2.4	5.2	5.5	6.8	16.1	—	—	11.3	6.7
Monoclinic pyroxene	0.1	1.1	R.g.	R.g.	4.3	—	R.g.	R.g.	0.45	R.g.	0.6	1.0	0.4	0.4	—	0.8	0.6	0.3	0.57	R.g.	2.0	R.g.	0.6	R.g.	1.5	1.3	0.5	R.g.	—	—	0.6	
Orthorhombic pyroxene	0.1	—	R.g.	0.1	—	R.g.	R.g.	0.6	0.8	0.3	R.g.	R.g.	1.5	0.5	—	R.g.	R.g.	0.6	R.g.	0.6	—	—	—	1.5	—	3.8	—	6.6	—	—	0.4	
Amphibole	2.5	4.2	4.4	3.2	0.9	2.1	10.4	2.7	9.0	19.4	9.9	3.9	2.8	3.9	3.3	0.9	3.9	2.7	3.0	5.3	5.04	0.7	2.9	0.5	6.1	6.6	1.5	—	—	—	4.2	4.0
Staurolite	0.1	R.g.	—	—	0.4	—	—	0.2	—	—	—	—	R.g.	—	—	0.3	—	—	—	—	—	—	—	R.g.	—	—	—	R.g.	—	—	0.04	
Sillimanite	—	—	—	—	—	—	0.3	—	—	—	—	—	—	—	—	—	—	—	—	—	—	—	—	R.g.	—	0.1	—	—	—	—	0.01	
Andalusite	—	—	—	—	—	—	—	0.2	—	—	—	—	—	—	—	0.2	—	—	0.1	—	—	—	—	—	—	—	—	—	—	—	0.01	
Kyanite	0.2	0.3	0.1	—	—	0.4	0.08	0.4	0.5	—	2.4	0.6	—	0.6	0.3	R.g.	0.27	0.2	0.44	0.4	—	—	—	—	—	0.4	—	—	—	—	0.27	
Zircon	36.3	20.5	32.0	60.1	22.6	17.2	21.9	41.4	20.5	10.58	23.0	8.0	23.2	13.8	30.2	0.4	2.4	1.5	31.8	23.5	32.66	18.7	15.9	30.0	23.2	13.3	18.0	13.7	—	—	24.4	13.6
Spinel	—	—	—	—	—	—	—	—	—	—	—	—	—	—	—	0.8	—	—	0.2	—	—	—	—	—	—	—	—	—	—	R.g.	0.04	
Chromite	10.7	7.2	3.2	11.9	3.8	—	10.4	16.6	9.0	7.4	9.1	3.8	17.6	11.4	10.7	0.4	4.9	18.2	17.6	4.9	9.8	9.2	11.4	8.3	2.2	8.4	2.5	17.5	—	—	8.8	5.2
Magnetite	—	—	4.7	1.8	1.7	33.3	—	0.1	0.1	5.0	0.7	1.2	1.2	2.4	2.5	8.3	R.g.	—	0.6	0.5	1.6	—	3.8	R.g.	R.g.	—	R.g.	0.5	—	—	2.4	6.2
Ilmenite	4.0	10.6	1.2	—	1.5	1.4	2.0	1.0	2.0	1.2	0.9	1.1	3.4	1.3	1.4	1.7	3.0	1.5	0.8	R.g.	2.25	9.8	2.3	1.2	1.5	1.1	2.0	2.9	—	—	2.2	2.4
Leucoxene	6.4	6.4	14.3	16.5	4.9	17.9	2.2	2.9	6.0	12.3	1.83	11.0	3.1	12.1	25.0	13.5	3.4	8.5	8.6	7.1	12.0	9.6	12.0	0.8	4.73	23.4	18.4	11.4	—	—	9.3	6.7
Anatase	0.2	3.2	1.0	0.4	0.7	—	0.6	—	R.g.	R.g.	0.3	0.1	—	—	R.g.	—	0.1	0.2	—	R.g.	—	—	—	—	—	0.4	—	—	—	—	0.3	
Nigrine	—	—	0.1	—	0.1	—	—	0.03	—	—	—	—	—	—	—	—	—	—	—	—	—	R.g.	—	—	—	—	—	—	—	—	0.01	
Rutile	6.2	7.2	3.1	2.8	3.7	3.4	6.7	4.1	1.3	0.6	0.4	0.7	2.8	2.3	4.0	0.5	0.5	4.1	3.4	8.92	4.9	9.0	1.87	0.54	3.6	—	1.3	3.3	—	—	3.2	2.6
Corundum	—	0.3	R.g.	R.g.	0.8	0.3	—	0.4	—	—	R.g.	0.4	—	—	14.8	4.2	0.2	0.3	0.8	0.29	0.9	0.2	—	—	—	—	0.3	—	—	—	0.8	
Hematite	0.2	3.4	1.4	R.g.	0.9	3.4	2.3	2.5	3.8	1.0	1.3	1.3	3.6	0.6	1.9	2.4	3.4	1.9	1.1	4.6	1.96	2.6	5.7	2.6	3.8	2.3	—	2.4	7.0	—	2.4	1.6
Limonite	0.9	4.0	0.3	—	0.6	—	2.2	—	1.6	0.9	3.3	1.1	1.8	1.9	—	87.0	0.5	—	—	14.28	2.3	6.0	26.0	5.0	2.5	—	—	2.4	3.8	—	5.7	16.5
Pyrite	—	—	—	—	—	—	—	—	—	—	—	—	—	—	—	R.g. 4.3	—	R.g.	—	—	0.1	—	—	—	—	—	—	—	—	R.g.	0.2	

*Here and below, R.g. means rare grains.

89

Minerals of the heavy fraction comprise not less than 0.7% of the rock, and their grain size varies from 0.01 to 0.7 mm. A list of all the minerals in the heavy fraction is given in Table 23. Brief descriptions of the typical ones are given below.

Zircon is represented by almost colorless crystals of dipyramidal habit with unrounded angles and edges. The length-width ratio of the crystal varies from 2:1 to 1.5:1. Many crystals contain inclusions of gas bubbles and dark opaque minerals. The other variety of zircon consists of larger prismatic crystals with well-developed prism faces. They contain many inclusions of large gas bubbles.

Epidote consists of greenish half-rounded grains, often in the form of aggregates with chlorite.

Ilmenite is one of the characteristic minerals of the heavy fraction of sandstones of the Bol'shoi Balkhan section. Its grains are large, tabular, slightly rounded, and always replaced by leucoxene to some extent.

Tourmaline forms crystal fragments up to 0.25 mm long. Their shape is so well preserved that crystals with faceted heads are found. The tourmalines are dark and pleochroic, from dark brown to light brown.

Rutile occurs in the form of prismatic crystals of a brilliant dark red color, slightly rounded, and in many cases the longitudinal striations on the faces are present.

Garnets consist of large pale pink grains of angular form. Many of them contain numerous inclusions of dark minerals. The latter may be magnetite, since the garnet grains containing inclusions are attracted by an iron needle.

Corundum is a characteristic mineral insofar as corundum grains attain 0.7 mm and are the largest grains in the nonmagnetic fraction. They are pale pink, irregularly distributed, sometimes in clusters, and optically uniaxial, $Ng = 1.767$, $Np = 1.754$.

The paragenetic associations of minerals from the Middle Jurassic section of Bol'shoi Balkhan can be established from the summary matrix (Table 24). An analysis of the matrix of correlation coefficients given in the table showed that there is a positive association between the following pairs of minerals: garnet–ilmenite, rutile–zircon, epidote–amphiboles, and quartz–feldspars.

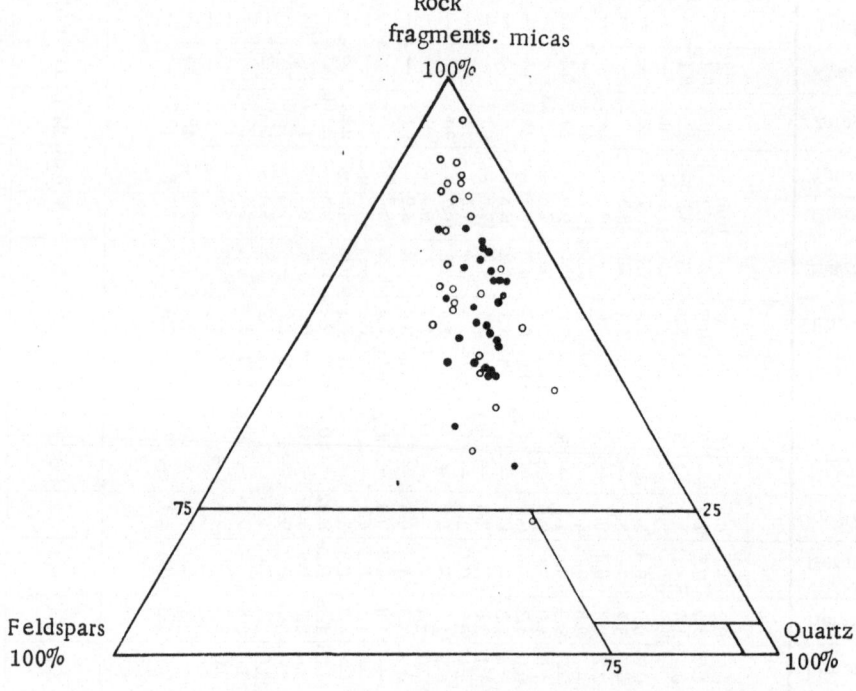

Fig. 70. Mineralogical composition of samples from Middle Jurassic deposits of West Turkmenia. Distribution of samples from Aman-Bulak section (circles) and from Balkhan section (filled-in circles) over field of classification triangle of Dapples, Krumbein, and Sloss.

The characteristic mineral associations are rutile–zircon and garnet–ilmenite.

The mean amount of soluble total iron in sandstones from the Bol'shoi Balkhan section is 4.18% with a standard deviation of 0.88%, and the ferrous iron content exceeds the ferric iron content (Table 25).

Middle Jurassic Deposits in Aman-Bulak Section (Tuarkyr Anticline)

The second investigated section of Middle Jurassic deposits in West Turkmenia lies in the region of the Tuarkyr anticline, where they are represented by deposits of platform facies.

TABLE 24. Matrix of Correlation Coefficients and Their Significances for Mineral Associations in Rocks from Middle Jurassic Deposits of West Turkmenia (Bol'shoi Balkhan Section)

Mineral	Ilmenite	Garnet	Magnetite	Chromite	Rutile	Zircon	Amphiboles	Epidote	Feldspars	Quartz	Leucoxene
Ilmenite		2.1	0.3	0.1	**2.1**	0.3	0.7	1.3	1.5	0.1	0.5
Garnet	**+0.40**		1.8	1.8	1.2	0.2	0.2	0.6	0.8	1.9	0.8
Magnetite	−0.05	−0.34		1.3	0.7	0.4	0.5	1.4	1.3	1.3	1.6
Chromite	−0.02	+0.34	−0.25		1.6	1.6	0.3	0.4	1.4	1.9	0.9
Rutile	**+0.40**	+0.23	−0.13	+0.30		**2.0**	0.6	0.5	0.1	1.4	0.3
Zircon	−0.06	+0.04	−0.08	+0.31	**+0.36**		0.5	0.6	1.2	0.2	0.5
Amphiboles.	−0.14	−0.03	−0.10	−0.05	−0.14	−0.10		**2.8**	1.4	0.7	0.5
Epidote.	+0.25	−0.11	−0.27	+0.07	−0.07	−0.11	**+0.50**		0.3	1.0	0.4
Feldspars	−0.29	−0.16	−0.25	+0.28	−0.01	+0.23	+0.28	+0.06		**2.5**	**2.2**
Quartz	+0.01	+0.36	−0.26	+0.36	+0.27	+0.04	+0.03	−0.20	**+0.45**		1.4
Leucoxene	0.10	+0.16	−0.31	+0.18	−0.05	−0.10	+0.10	+0.10	**+0.41**	+0.27	

Note: The lower left half of the table gives the values of the correlation coefficients r; the upper right half gives the significances t of the correlation coefficients. Bold figures denote significant positive r.

The Middle Jurassic deposits comprising the arch of the Tuarkyr anticline were investigated in the region of the Aman-Bulak well, where beds of grayish gritstones, sandstones, siltstones, and clays with bands of carbonaceous shales are exposed over an extensive area. According to Kurbatov (1956), the Middle Jurassic deposits here are subdivided into two parts: a lower, predominantly carbonaceous-clayey part, and an upper, predominantly sandy part of marine origin.

The investigated section begins south of the Aman-Bulak well and runs in an eastern direction. The units of the section are described from the bottom upward:

1. The deposits exposed at the base of the section are fine-pebble conglomerates of varying thickness, reaching 30-40 m in places.

2. The overlying unit consists of gray compact sandstones alternating with bands of silty clays. The thickness of the unit reaches 40 m.

3. A predominantly clayey unit with beds of gray sandstones and relatively thick coal seams. Thickness of unit 65 m.

4. Clayey sandstones and sandy clays with separate thick beds of compact gray sandstones. The top of the unit is composed of compact brownish gray sandy clays. Thickness 135 m.

5. Unit of gray sandstones and dark silty clays with interlayers of coal. The upper coal seam contains bands of opokas and opoka-like (opaline) clays.

TABLE 25. Results of Chemical Analyses of Rocks from Middle Jurassic Deposits of West Turkmenia

Aman-Bulak section				Bol'shoi Balkhan section			
No. of sample	Total iron	Fe_2O_3	FeO	No. of sample	Total iron	Fe_2O_3	FeO
1 c	1.98	1.71	0.27	1	4.27	1.27	2.65
1 b	2.65	1.07	1.42	1	4.58	1.01	3.22
1 a	2.40	1.26	1.03	3	4.88	1.41	3.13
1	2.51	1.95	0.50	4	4.39	0.65	3.37
2	1.87	0.25	1.46	5	5.16	1.52	3.28
3	1.43	1.24	0.17	6	5.19	2.05	2.83
4	3.36	3.30	0.15	7	4.05	1.09	2.67
5	1.32	0.40	0.83	8	3.85	1.47	2.14
6	3.24	2.40	0.76	9	3.84	1.20	2.38
7	4.25	3.08	1.06	10	6.02	1.55	4.03
8	4.22	3.01	1.09	11	4.39	1.05	3.01
9	0.49	0.30	0.17	12	4.90	2.92	1.78
10	4.43	3.87	0.50	13	4.31	1.50	2.53
11	8.70	8.31	0.35	14	4.67	2.18	2.24
12	6.24	5.30	0.85	15	3.98	2.03	1.76
13	4.39	3.04	1.22	16	4.36	2.03	2.10
14	2.02	1.03	0.65	17	4.15	2.00	1.94
15	1.56	1.10	0.42	18	2.87	1.12	1.58
16	9.59	6.57	2.72	19	3.46	0.90	2.31
17	1.55	0.77	0.70	20	2.72	1.48	1.12
18	6.21	6.08	0.12	21	3.47	1.19	2.05
19	0.58	0.30	0.25	22	4.83	3.10	1.56
20	5.02	2.37	2.39	23	4.59	1.69	2.61
21	17.38	17.29	0.09	24	3.85	2.81	0.94
22	4.41	3.85	0.50	25	5.75	5.22	0.48
23	4.22	1.96	2.04	26	2.66	1.58	0.97
24	1.53	1.06	0.42	28	4.56	3.73	0.75
25	0.34	0.31	0.03	29	3.37	1.62	1.58
				30	1.99	1.30	0.62
$\bar{X}$	3.85	2.98	0.79	$\bar{X}$	4.18	1.82	2.13
S	3.45	3.42	0.69	S	0.88	0.94	0.88

6. Unit of compact thick-bedded sandstones, sometimes coarse-grained and containing clay balls. Thickness 46 m.

7. Predominantly clayey unit with beds of gray sandstones and thin coal seams. Thickness of unit 66 m.

The total thickness of the described Middle Jurassic deposits is 529 m. These are overlain by Middle Jurassic deposits consisting mainly of similar beds of gray or brownish gray sandstones (marine Middle Jurassic).

8. Compact medium- and coarse-grained sandstones of a brownish gray color. In places at the base of the unit there are pebble-bearing conglomerates in the sandstones of the Middle Jurassic coal-bearing strata and ferruginized fragments of wood, which reach a large size. Thickness 27 m.

9. Medium-grained gray sandstones containing bun-shaped or spherical concretions. Thickness 14.5 m.

10. Unit of clays, thin-layered, carbonaceous in places. Thickness 8 m.

11. Unit of gray compact sandstones alternating with fine-grained sandstones. The upper part of the unit contains a bed of calcareous clays and marls with remains of a marine fauna: Macrodon sp., Astarte cf. pulla Roem., Pseudomonotis sp., and Modiola gigantea Quenst. Thickness 37 m.

12. Predominantly clay unit with rare beds of gray sandstones containing large calcareous sandy concretions with an oyster fauna. Thickness 19.5 m.

13. Clays, sandy clays, and sandstones. The sandstone beds contain spherical concretions with a fauna. Thickness 8 m.

TABLE 26. Granulometric Composition and Carbonate Content of Samples from Middle Jurassic Deposits in Aman-Bulak Section (West Turkmenia)

| No. of sample | Carbonate content | Fractions, mm | | | | | | | | | | Total |
		> 0.59	0.59—0.42	0.42—0.29	0.29—0.21	0.21—0.149	0.149—0.105	0.105—0.074	0.074—0.053	0.053—0.01	< 0.01	
1c	31.9	0.3	0.3	7.9	23.7	33.2	12.0	2.7	3.4	2.0	14.4	99.9
1b	42.5	0.1	—	0.1	0.2	2.6	9.6	22.9	24.5	10.4	29.6	100.0
1a	44.3	—	—	0.3	0.9	11.3	49.0	15.6	8.6	3.7	10.4	99.8
1	1.8	3.3	5.7	18.5	18.2	21.4	5.8	1.6	3.1	2.6	19.0	99.2
2	46.3	—	0.1	18.8	42.2	20.4	6.5	1.2	2.5	2.1	6.5	100.3
3	3.8	9.1	17.5	26.4	8.6	11.5	5.8	2.1	2.8	2.5	13.5	99.8
4	17.0	0.7	0.3	3.2	15.6	43.5	9.6	2.4	3.2	2.6	18.7	99.8
5	30.7	—	0.3	4.5	23.1	44.3	8.4	2.4	2.8	2.2	11.9	99.9
6	34.0	—	—	—	—	0.1	1.1	5.1	47.9	20.8	24.7	99.8
7	54.5	—	—	—	—	5.8	37.1	9.1	9.0	7.6	31.1	99.6
8	54.5	—	—	—	—	—	10.9	20.8	19.0	10.2	38.7	99.6
9	33.0	—	—	0.9	17.6	64.3	7.2	1.5	1.7	1.1	5.4	99.7
10	40.4	—	—	—	0.2	1.0	15.5	17.3	19.2	10.4	36.2	99.8
11	53.9	—	—	—	—	0.5	15.1	12.1	11.5	9.6	51.0	99.8
12	37.4	—	—	—	0.3	3.0	31.7	11.2	9.5	9.9	34.4	100.0
13	26.8	—	—	—	0.1	0.6	15.4	18.3	20.1	10.0	35.1	99.6
14	26.8	—	—	—	0.1	0.5	2.1	7.3	40.2	18.1	31.5	99.8
15	1.2	0.03	0.03	0.1	1.0	42.7	19.1	3.3	5.2	4.0	24.5	100.0
16	39.3	—	—	—	—	0.1	1.6	5.8	21.6	22.4	48.3	99.8
17	71.6	—	0.1	0.2	3.7	31.1	14.1	3.4	4.8	4.2	38.0	99.6
18	29.6	—	0.1	0.2	4.5	29.6	17.6	4.6	5.9	6.3	31.1	99.9
19	23.7	1.3	4.0	24.7	24.5	14.6	7.7	2.6	3.4	3.3	13.5	99.6
20	32.7	1.0	2.4	21.4	20.8	12.0	9.7	5.6	5.9	4.9	16.2	99.9
21	25.2	—	—	—	2.9	21.7	12.3	4.0	5.5	8.2	45.3	99.9
22	19.6	5.8	2.6	19.2	21.1	14.0	6.4	2.4	4.3	4.9	19.3	100.0
23	38.5	—	0.2	7.2	14.6	28.2	16.1	5.5	6.3	5.2	16.6	99.9
24	24.9	—	0.2	1.8	12.6	38.9	13.6	3.6	5.3	5.0	19.0	100.0
25	0.8	1.2	7.6	61.3	13.8	4.7	2.1	0.8	1.1	1.0	6.4	100.0
$\overline{X}$	30.7					17.9	13.0	7.0	10.7	7.0	24.6	
S	18.5					17.2	10.4	6.4	11.6	5.7	13.0	

14. Brownish clayey sandstones, pale brown massive sandstones, and beds of gray platy sandstones. Thickness 29 m.

The total thickness of the Middle Jurassic marine deposits is 143 m.

The total thickness of the whole Middle Jurassic deposits in the Aman-Bulak section is 672 m.

For photometric measurements we took samples only from layers of sandy or silty composition so that we could characterize the considered section as uniformly as possible. The total extent of the section on the surface of the outcrops was about 10 km. We photometered the surface of the samples in their natural state, free from any authigenic alteration. The total number of samples from the Aman-Bulak section was 28.

As the results of laboratory analyses showed, the investigated samples are silty-sand rocks, mainly sandstones (Table 26). The positions of the points of these samples on the classification triangle (see Fig. 69) show that 16 of them are sands, three are silts, one a mudstone, and seven samples are rocks of intermediate composition. The carbonate content of the rocks is extremely variable – from 1.75 to 71.6%. The mean content is 31.65% and the standard deviation S = 16.86%. These values of the carbonate content include the carbonate content of the cement and the fragmentary particles.

The microstructure of the rocks is sandy or silt-sandy with basal or poikiloclastic, sometimes interstitial, cement. The mean cement content is 46.1%. Carbonate cement is most common. Limonite cement is rarer.

TABLE 27. Petrographic Composition of Rocks from Central Jurassic Deposits in Aman-Bulak Section (West Turkmenia)

No. of sample	Grain size, mm	Content of fragments, % (by vol.)				
		feldspars	quartz	mica	rock fragments	cement
1 c	0.02—0.5	10.7	33.2	0.4	13.3	42.4
1 b	0.01—0.1	9.2	9.0	1.0	12.6	68.2
1 a	0.07—0.2	13.7	13.1	0.4	20.8	52.0
1	0.10—0.80	20.7	30.4	4.7	37.3	6.9
2	0.10—0.70	7.8	24.9	0.2	23.8	43.3
3	0.10—0.35	29.6	36.3	0.6	24.0	9.5
4	0.05—0.50	22.3	31.6	1.8	32.3	12.0
5	0.05—0.50	17.0	16.7	—	30.7	35.6
6	0.01—0.07	6.5	10.4	—	7.7	75.4
7	0.01—0.10	8.6	5.9	—	5.0	80.5
8	0.05—0.50	7.5	10.1	—	15.8	66.6
9	0.10—0.50	12.9	43.2	—	15.8	28.1
10	0.10—0.80	11.7	9.7	—	36.0	42.6
11	0.05—0.50	5.3	9.3	—	15.1	70.3
12	0.05—0.22	7.5	13.9	—	25.6	53.0
13	0.01—0.10	8.3	15.6	—	32.9	43.2
14	0.01—0.10	5.4	12.9	2.8	22.8	56.5
15	0.02—0.30	19.8	29.4	0.9	48.6	1.3
16	0.01—0.07	2.0	5.0	—	3.5	89.2
17	—	—	—	—	—	100.0
18	0.01—0.80	14.4	23.2	—	11.9	50.5
19	0.10—0.70	18.7	21.1	—	32.7	27.5
20	0.05—0.50	19.9	17.1	—	11.7	51.3
21	0.07—0.50	19.7	21.1	—	19.9	39.3
22	0.10—1.5	24.1	19.8	—	26.0	30.1
23	0.1—0.7	19.2	17.5	—	15.5	47.8
24	0.01—0.10	10.0	11.3	—	11.8	66.9
25	0.10—0.40	25.3	52.1	0.5	22.1	—
$\overline{X}$		13.5	19.4	0.5	20.5	46.07
S		7.4	11.9	1.0	11.2	26.18

TABLE 28. Mineralogical Composition of Heavy Fraction of Samples from Middle Jurassic Deposits in Aman-Bulak Section (West Turkmenia)

No. of sample	Pyrite	Limonite	Hematite	Corundum	Rutile	Nigrine	Anatase	Leucoxene	Ilmenite	Magnetite	Chromite	Zircon	Kyanite	Sillimanite	Staurolite	Amphibole	Orthorhombic pyroxene	Monoclinic pyroxene	Epidote	Garnet	Tourmaline	Biotite	Chlorite	Sphene	Monazite	Apatite	Barite	Carbonates	Rock fragments
1 c	Rg	3.8	15.1	Rg	0.8	—	—	1.4	7.0	1.5	3.2	8.4	0.1	—	1.3	1.9	0.1	0.6	1.9	19.6	Rg	—	1.0	—	Rg	0.5	—	0.1	31.7
1 b	—	5.2	19.6	—	2.2	—	Rg	—	16.8	3.0	18.8	8.6	—	—	—	4.0	0.3	0.6	1.7	3.8	0.3	—	0.3	—	—	4.0	—	0.3	10.5
1 a	—	2.7	32.3	Rg	0.6	—	—	0.23	21.7	14.5	4.0	9.23	0.1	—	—	1.4	0.07	Rg	4.0	0.7	0.5	0.3	0.3	—	Rg	0.2	—	0.07	7.1
1	—	27.2	4.6	1.9	1.0	—	—	0.5	8.5	Rg	0.5	8.4	0.2	—	Rg	0.8	0.1	0.3	1.6	35.5	0.8	0.1	Rg	0.1	—	0.4	—	0.3	7.2
2	0.1	2.7	17.7	0.2	Rg	—	—	1.0	29.2	Rg	3.2	6.4	0.2	—	Rg	1.9	0.2	1.1	7.2	13.7	0.2	—	Rg	—	—	0.6	1.1	0.1	14.3
3	—	33.4	11.5	0.6	0.5	—	—	1.8	21.8	2.8	1.2	5.7	0.5	—	—	1.2	—	—	3.1	6.2	Rg	0.3	0.3	0.5	—	1.5	0.3	2.7	3.8
4	0.8	30.5	7.4	0.8	1.0	—	—	2.2	14.0	Rg	1.2	2.4	0.4	—	0.3	4.1	1.0	1.2	2.1	14.4	1.7	0.4	0.8	—	0.4	1.2	0.1	0.4	10.8
5	—	1.1	15.0	—	0.6	—	—	1.8	47.7	4.2	3.4	6.7	0.1	—	—	3.5	—	—	1.8	9.3	Rg	0.7	0.2	—	—	0.3	Rg	0.1	3.1
6	—	64.2	4.2	—	—	—	—	1.6	8.4	3.7	Rg	—	—	—	0.3	9.9	—	0.6	5.3	1.1	—	1.1	—	—	—	—	—	0.5	—
7	1.0	9.0	14.5	—	—	—	—	2.6	32.2	5.3	0.3	2.9	Rg	—	—	8.7	1.0	0.6	6.1	8.4	0.3	Rg	0.6	Rg	—	1.3	—	Rg	6.8
8	—	1.5	13.5	1.2	0.1	—	—	0.5	9.1	1.2	0.3	1.1	0.2	—	—	5.7	0.9	—	14.3	45.4	0.3	Rg	1.0	0.3	0.2	0.2	Rg	Rg	1.8
9	—	8.3	6.0	—	1.8	—	0.5	17.5	10.0	5.9	Rg	5.0	1.0	—	—	6.2	0.5	Rg	4.0	1.8	0.2	—	0.5	—	—	1.0	—	—	29.0
10	0.01	63.0	9.5	—	Rg	—	—	0.1	1.1	1.1	1.2	0.5	0.1	—	—	1.1	Rg	—	0.5	Rg	—	0.05	0.3	Rg	—	0.05	—	0.1	22.3
11	—	67.3	9.8	—	0.3	Rg	Rg	0.9	3.1	0.4	Rg	0.7	Rg	—	—	Rg	—	Rg	0.03	0.3	Rg	—	—	—	—	Rg	—	15.4	—
12	Rg	86.2	4.2	—	0.9	—	0.01	Rg	1.3	0.7	1.9	1.8	0.05	—	—	0.3	Rg	Rg	1.03	2.6	Rg	0.3	0.2	Rg	—	0.1	—	0.2	0.3
13	—	20.5	17.7	—	0.9	0.1	Rg	2.1	20.5	6.7	0.3	9.6	0.5	0.2	—	6.8	Rg	Rg	3.3	7.6	Rg	Rg	—	—	—	0.6	—	—	1.0
14	—	81.3	1.9	Rg	0.2	—	—	0.9	1.4	3.4	2.5	2.7	0.3	—	0.4	1.0	0.2	Rg	3.3	Rg	—	—	0.2	Rg	—	0.4	—	0.7	2.0
15	—	1.7	13.5	—	2.7	—	—	7.0	35.2	6.7	0.6	2.1	—	0.1	—	0.2	0.2	0.3	7.3	13.5	Rg	—	—	—	—	0.6	—	—	6.6
16	—	59.7	15.6	1.1	—	—	—	5.3	5.2	2.7	1.5	1.9	0.6	—	—	2.7	Rg	0.7	2.2	4.7	0.7	0.1	1.1	Rg	—	1.1	—	1.6	—
17	—	10.4	8.5	—	0.5	—	Rg	0.2	7.9	Rg	—	9.0	Rg	—	0.2	1.1	Rg	—	4.1	37.1	0.2	—	0.7	Rg	—	0.1	—	0.2	10.3
18	—	94.3	—	—	0.6	Rg	Rg	1.7	0.5	Rg	2.1	2.4	0.1	—	—	Rg	0.4	0.2	0.5	—	—	—	—	Rg	—	0.2	—	1.0	0.1
19	—	8.7	13.4	—	0.7	—	0.6	0.5	13.6	Rg	—	12.8	Rg	—	—	3.3	—	—	7.1	2.8	Rg	—	0.7	Rg	—	0.4	—	1.7	29.4
20	—	12.1	72.5	Rg	0.2	—	—	—	10.9	0.5	—	0.9	Rg	—	—	0.2	0.5	0.2	1.0	1.2	0.2	—	—	Rg	—	Rg	—	Rg	Rg
21	—	99.8	—	—	Rg	Rg	—	0.7	—	Rg	0.7	0.2	Rg	—	—	Rg	—	Rg	—	Rg	—	—	—	—	—	Rg	—	Rg	Rg
22	—	59.2	9.7	—	0.5	—	—	—	0.9	0.2	0.6	0.9	Rg	—	—	0.9	Rg	Rg	1.8	13.7	—	1.1	0.2	Rg	—	Rg	—	Rg	9.6
23	Rg	36.0	10.4	—	2.8	—	—	1.7	17.8	Rg	—	8.6	Rg	—	—	0.2	Rg	Rg	0.7	2.1	—	—	0.5	0.1	Rg	Rg	—	0.1	20.0
24	—	21.1	5.5	—	1.1	—	—	—	6.8	0.9	—	22.4	0.1	—	—	1.9	Rg	—	3.8	2.3	Rg	—	1.1	0.3	—	—	—	0.8	29.6
25	—	11.4	6.1	—	Rg	0.1	—	—	3.4	Rg	—	1.8	0.1	—	—	1.3	—	—	3.0	1.9	—	1.1	—	Rg	0.4	Rg	—	66.7	2.7
X̄		32.9	12.8						1.9	12.7	2.3						2.5			3.3	8.9						0.5		3.3	9.3
S		32.0	7.2						3.5	9.9	3.2						2.7			3.1	12.4						0.8		12.8	10.4

95

The composition and grain size of the minerals of the light fraction are given in Table 27 and illustrated in the form of points on a classification triangle (Fig. 70). The investigated samples are typical graywackes in composition. Only the sample taken from the top of the Bathonian stage is close to arkose.

The rock-forming minerals of the considered samples are characterized by the following features:

Quartz occurs in the form of angular grains containing gas-bubble inclusions. Many of the grains exhibit mosaic extinction and the corrosion of quartz grains by the carbonate cement is very common.

The feldspars consist mainly of perthitic microclines with distinct grid twinning. Plagioclases occasionally occur in the form of angular grains of albite and albite-oligoclase.

TABLE 29. Matrix of Correlation Coefficients and Their Significances for Mineral Associations in Rocks from Middle Jurassic Deposits in Aman-Bulak Section (West Turkmenia)

Mineral	Hematite	Ilmenite	Magnetite	Chromite	Zircon	Amphiboles	Epidote	Feldspars	Quartz	Rock fragments	Carbonates	Limonite
Hematite	—	2.8	1.6	2.9	1.1	0.3	1.0	0.2	1.0	1.6	1.7	3.6
Ilmenite	**+0.52**	—	3.5	3.6	1.4	1.3	2.2	0.6	0.0	1.1	0.2	3.4
Magnetite	+0.32	+0.60	—	1.3	0.3	2.6	1.6	0.8	0.3	0.5	0.4	1.6
Chromite	**+0.53**	**+0.62**	+0.26	—	3.6	0.6	1.4	0.7	0.7	0.5	1.4	3.1
Zircon	+0.22	+0.27	+0.05	**+0.62**	—	0.1	0.7	0.1	0.1	0.3	1.5	2.7
Amphiboles	+0.05	+0.26	**+0.47**	+0.12	+0.02	—	3.2	1.3	0.2	1.2	1.5	1.2
Epidote	+0.20	**+0.41**	+0.32	+0.27	+0.13	**+0.56**	—	1.0	0.1	0.1	1.0	2.7
Feldspars	-0.04	+0.12	-0.15	-0.14	-0.03	-0.26	-0.19	—	4.0	1.4	3.9	0.6
Quartz	-0.20	0.00	-0.05	-0.13	-0.01	-0.03	-0.02	**+0.67**	—	4.0	4.0	1.1
Rock fragments	-0.01	+0.21	-0.10	+0.10	-0.06	-0.23	-0.01	+0.28	+0.34	—	3.9	0.3
Carbonates	+0.34	+0.03	+0.08	+0.27	+0.29	+0.30	+0.19	-0.65	-0.67	-0.65	—	1.4
Limonite	-0.62	-0.59	-0.32	-0.56	-0.50	-0.23	-0.50	-0.12	-0.20	-0.06	-0.27	—

Note. The lower left half of the table gives the values of the correlation coefficients r; the upper right half gives the significances t of the correlation coefficients. Bold figures denote significant positive and negative r.

96

All the considered samples from the Aman-Bulak section show corrosion of the feldspar grains by carbonate cement.

Mica occurs in increased amount in samples of fine-grained sandstones and in siltstones and consists of small greenish scales.

Rock fragments occur in appreciably smaller quantity in samples from the Aman-Bulak section than in samples from the Bol'shoi Balkhan section, and their quantity varies (Tables 22 and 27). The rock fragments include many medium-grained granites and quartz porphyries. Granite porphyries and granophyric granites are much rarer, and fragments of sandstones, siltstones, and clayey rocks are sometimes present.

The composition of the minerals of the heavy fraction is shown in Table 28. It should be noted that the heavy fraction contains a high content of limonite, which is usually mixed with carbonates, and hematite, which forms flat grains with a rough surface and a film of ferric hydroxide.

Ilmenite occurs in thickly tabular grains with well-preserved faces and a strong metallic luster. In samples from the upper part of the section the ilmenite grains are more rounded and some are replaced by leucoxene.

The presence of minerals from metamorphic rocks (garnets and kyanite) is characteristic. The grains of these minerals are the largest among the minerals of the heavy fraction. Individual grains of kyanite reach a length of 0.6 mm. Some of them are bent and twisted.

The paragenetic relations of the minerals of the investigated samples from the Aman-Bulak section are shown in Table 29, which gives the correlation coefficients and their significances after transformation by Fisher's method.

An analysis of the matrix in Table 29 shows that almost all the ore minerals (hematite, ilmenite, magnetite, and chromite) are associated with one another. The components of the group of opaque minerals are positively associated with epidote and amphibole.

Minerals of metamorphic rocks (kyanite and garnet) are not associated with other minerals.

The strong negative relationship between limonite and carbonates cannot be explained, since the content of these minerals is very high and the mineral associations have no constant component (Vistelius, Sarmanov, 1961. .

Chemical analyses for the content of soluble iron in the samples from the Aman-Bulak section showed that the total iron content of the samples is a little lower than in samples from the Bol'shoi Balkhan section (see Table 25), but the iron content is extremely variable and consists mainly of ferric iron.

Thus, the samples of sandstones and siltstones taken from Middle Jurassic sections formed in different conditions of sedimentation and difficult to distinguish from one another visually have their own petrographic characteristics. The comparison of mineral associations of samples from the Bol'shoi Balkhan and Aman-Bulak sections showed that:

1. Rocks from the platform section are characterized by an ilmenite – chromite association. The Bol'shoi Balkhan section, the rocks of which were formed in conditions close to geosynclinal, is characterized by rutile – zircon and ilmenite – garnet associations.

2. Rock samples from the platform section are much richer in ferric iron, while rocks from the geosynclinal section are richer in ferrous iron.

Results of Photometry of Samples from Middle Jurassic Sections

The results of photometry of the investigated samples are given in Table 30 in the form of the mean $\bar{\rho}_\lambda$ within the color thresholds and in the form of the polynomial coefficients of the spectral luminance curves (Tables 31, 32).

The appearance of the ρ_λ curves is illustrated in Fig. 71, where A denotes the curves of samples from the

TABLE 30. Mean Spectral Luminance Factors within Color Thresholds for Rocks
of Compared Middle Jurassic Sections of West Turkmenia

No. of sample	Aman Bulak section $\bar{Q}_{420-490}$	$\bar{Q}_{490-570}$	$\bar{Q}_{570-680}$	No. of sample	Bol'shoi Balkhan section $\bar{Q}_{420-490}$	$\bar{Q}_{490-570}$	$\bar{Q}_{570-680}$
1c	18.8	23.4	27.7	1	17.4	22.4	29.0
1b	20.5	25.7	31.9	2	15.7	19.3	23.9
1a	16.2	20.9	29.5	3	14.6	19.3	24.3
1	17.3	23.0	32.0	4	16.5	20.8	24.4
2	13.6	16.2	19.8	5	11.0	14.8	19.3
3	18.0	24.4	34.6	6	12.5	16.2	20.6
4	12.5	15.4	24.4	7	14.4	17.7	22.2
5	18.1	22.7	26.8	8	16.3	21.2	28.2
6	19.3	25.3	31.0	9	15.6	20.7	26.2
7	16.5	21.0	26.3	10	12.8	16.2	20.5
8	11.1	14.6	21.0	11	17.6	21.9	26.0
9	20.7	25.5	32.1	12	14.8	19.8	26.0
10	11.6	16.2	23.6	13	15.1	19.3	25.0
11	16.2	20.6	28.1	14	12.4	15.7	20.0
12	16.3	21.8	29.7	15	15.9	20.1	25.1
13	18.1	23.9	31.4	16	14.7	18.8	24.4
14	25.1	32.8	41.4	17	12.7	16.7	21.3
15	25.2	31.9	41.2	18	12.9	16.1	21.1
16	22.1	28.1	38.0	19	13.2	16.7	20.9
17	21.2	26.7	33.1	20	13.6	17.6	23.4
18	18.6	22.7	30.9	21	16.6	20.5	25.3
19	18.3	22.3	27.0	22	13.5	18.0	23.5
20	17.6	22.0	28.7	23	20.0	25.2	29.6
21	14.5	18.0	26.2	24	15.9	19.6	25.5
22	15.1	19.2	25.6	25	14.0	18.1	23.5
23	14.7	19.2	26.7	26	16.9	22.3	28.4
24	20.7	26.1	34.0	28	12.4	18.2	26.5
25	17.9	22.7	29.6	29	13.2	17.2	22.5
				30	19.4	24.7	35.1
$\overline{X}$	17.7	22.6	29.7		14.9	19.1	24.5
S	3.4	4.4	5.1		2.2	2.6	3.4

Aman-Bulak section, and B denotes those from the Bol'shoi Balkhan section. Externally the curves are difficult to distinguish from one another and a test of the differences in the curves from the mean ρ_λ within the color thresholds by the χ^2 method showed that this difference is insignificant, since the confidence level of the difference did not allow rejection of the uniform null hypothesis.

A difference in the nature of the ρ_λ curves of samples from the compared sections was revealed by a comparison of the polynomial coefficients of these curves. The values of ρ_λ obtained by photometry were smoothed by the above-described method (p. 116) and the polynomial coefficients up to the third power were calculated for the samples from each section (see Tables 31 and 32). The frequency distribution of the polynomial coefficients of the ρ_λ curves for the Middle Jurassic deposits of West Turkmenia is shown in Fig. 72.

A comparison of the values of the polynomial coefficients by the χ^2 method enabled us to establish the following relationships between these values and the region (Table 33) from which the investigated sample was taken.

From Table 33 we find $\chi^2 = 8.4$; i.e., the reliability of the relationship between a_1 and the source of the sample is more than 99.5%. Hence with a sufficiently large number of observations we can indicate the source of the sandstone sample or determine the lithofacies characteristics of the photometered rock. The reliability of the relationship between the value of a_2 and the source of the photometered sample lies between 98 and 99%

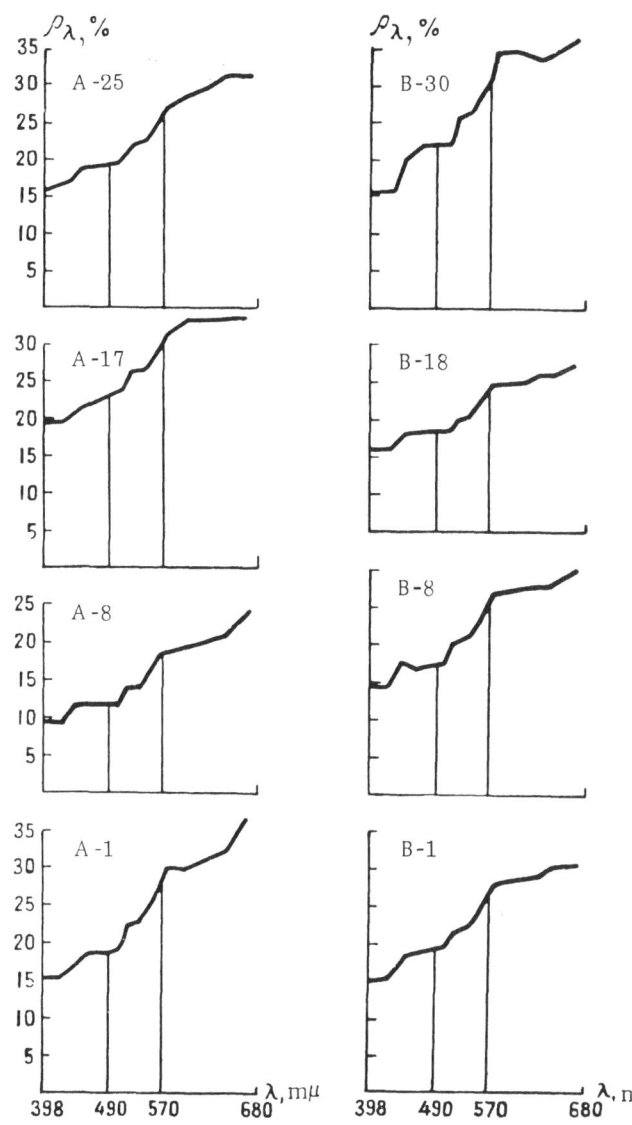

Fig. 71. Nature of ρ_λ curves of some rock samples from Middle Jurassic deposits of West Turkmenia.

(Table 34). The reliability of the relationship between the values of a_3 and the source of the sample lies between 80 and 90% (Table 35).

Thus, the data of a comparison of sections from the values of the polynomial coefficients of the spectral luminance curves of the investigated Middle Jurassic sandstones formed in different conditions of sedimentation showed a significant difference in the ρ_λ curves for samples from different regions. Hence an analysis of the measured values of ρ_λ showed that the types of spectral luminance curves of sandy rocks of the investigated Middle Jurassic sections are different.

The association between the petrographic-mineralogical composition of an investigated rock and its spectral luminance was determined by means of correlation methods.

Correlation Between Spectral Luminance of Rock and Its Composition

To investigate the effect of the composition of the rock on its spectral luminance, we calculated more than a hundred paired correlation coefficients. We calculated the correlation coefficients between the values of the mean ρ_λ within the color thresholds and the rock components, and between the values of the polynomial coefficients of the ρ_λ curves and the same rock components. Most of the calculated correlation coefficients were insignificant. Some of the calculated correlation coefficients were significant. From these we were able to determine the relationship between the composition of the rock and its spectral luminance.

Table 36 lists the values of r (numerator) and their significance t (denominator) evaluated by the use of Fisher's transformation. From an analysis of the correlation coefficients given in Table 36 we can draw the following conclusions:

1. The value of the mean spectral luminance within the color thresholds is affected by:

a) the content of soluble ferric iron in the rock, which reduces the values of $\overline{\rho}_\lambda$ in the short-wave part of the spectrum (Aman-Bulak section);

b) the content of soluble ferrous iron, which reduces the values of $\overline{\rho}_\lambda$ in the short-wave part of the spectrum (Bol'shoi Balkhan section);

c) all the mean $\overline{\rho}_\lambda$ increase with increase in the amount of quartz fragments in the rock;

d) an increase in the carbonate content of the rocks reduces the spectral luminance in the medium- and long-wave parts of the spectrum;

e) an increase in the clay content reduces the values of $\overline{\rho}_\lambda$ in the short- and long-wave parts of the spectrum.

TABLE 31. Values of Polynomial Coefficients of ρ_λ Curves of Rocks from Middle Jurassic Deposits of West Turkmenia (Bol'shoi Balkhan Section)

No. of sample	a_0	a_1	a_2	a_3
1	15.0178	−0.1287	+0.2621	−0.0125
2	13.9282	−0.0931	+0.1884	−0.0093
3	12.3493	−0.0166	+0.2084	−0.0102
4	13.6262	+0.5780	+0.1133	−0.0070
5	9.4442	−0.3106	+0.2390	−0.0118
7	10.4648	+0.5143	+0.1171	−0.0062
8	15.9641	−1.1380	+0.3803	−0.0163
9	14.8295	−0.9889	+0.3925	−0.0188
10	11.2834	−0.1735	+0.1923	−0.0092
11	16.2871	−0.2543	+0.2193	−0.0105
12	14.2451	−1.1896	+0.4206	−0.0195
13	15.5578	−1.4533	+0.4111	−0.0183
14	10.7927	−0.2915	+0.2295	−0.0115
15	15.2632	−0.8091	+0.3120	−0.0143
16	13.6932	−0.6293	+0.2903	−0.0133
17	11.0350	−0.1980	+0.2256	−0.0112
18	11.7968	−0.3758	+0.2092	−0.0092
19	11.2132	+0.1646	+0.1445	−0.0073
20	11.6349	−0.1750	+0.2360	−0.0102
21	14.9397	−0.1526	+0.2078	−0.0100
22	10.5084	+0.4092	+0.1500	−0.0075
23	16.6325	+0.0115	+0.2697	−0.0143
24	15.4388	−0.9653	+0.3329	−0.0149
25	13.3633	−0.0364	+0.1374	−0.0050
26	14.4221	−0.1185	+0.2711	−0.0132
28	12.9307	−1.9608	+0.5892	−0.0272
29	11.7087	−1.0012	+0.3509	−0.0163
30	15.3361	+0.0701	+0.3562	−0.0183
$\overline{X}$	13.3467	−0.3826	0.2663	0.0126
S	2.0608	0.6049	0.1088	0.0049

TABLE 32. Values of Polynomial Coefficients of ρ_λ Curves of Rocks from Middle Jurassic Deposits of West Turkmenia (Aman-Bulak Section)

No. of sample	a_0	a_1	a_2	a_3
1c	17.6002	−0.5555	0.2955	−0.0145
1b	18.4617	−0.0072	0.2078	−0.0090
1a	17.7666	−2.2158	0.5507	−0.0233
1	15.7132	−0.4997	0.3012	−0.0115
2	14.1192	−1.0484	0.2742	−0.0120
3	14.2078	+0.5176	0.2118	−0.0092
4	13.8334	−1.8907	0.4232	−0.0156
5	15.8449	+0.2545	0.1605	−0.0085
6	15.1942	+0.6511	0.1927	−0.0110
7	15.4330	−0.7045	0.3230	−0.0153
8	10.1731	−0.4659	0.2125	−0.0078
9	18.7052	−0.2359	0.2553	−0.0115
10	11.1911	−1.2223	0.4033	−0.0175
11	15.6671	−1.0124	0.3520	−0.0047
12	14.0031	−0.0976	0.2436	−0.0100
13	17.8772	−1.4797	0.4853	−0.0217
14	22.4842	−0.5570	0.4348	−0.0208
15	21.5648	−0.1158	0.3513	−0.0166
16	21.7744	−1.6905	0.5351	−0.0228
17	20.6217	−1.1471	0.4301	−0.0202
18	19.1250	−1.6420	0.4368	−0.0177
19	16.5663	−0.2388	0.2242	−0.0107
20	16.7137	−0.8637	0.3259	−0.0137
21	15.2950	−1.6121	0.3899	−0.0145
22	14.2973	−0.8411	0.3125	−0.0132
23	14.4476	−1.3428	0.4225	−0.0183
24	20.4561	−1.3031	0.4299	−0.0185
25	17.3322	−1.2061	0.4055	−0.0177
$\overline{X}$	16.6596	−0.8062	0.3425	−0.0149
S	3.0420	0.7315	0.1064	0.0045

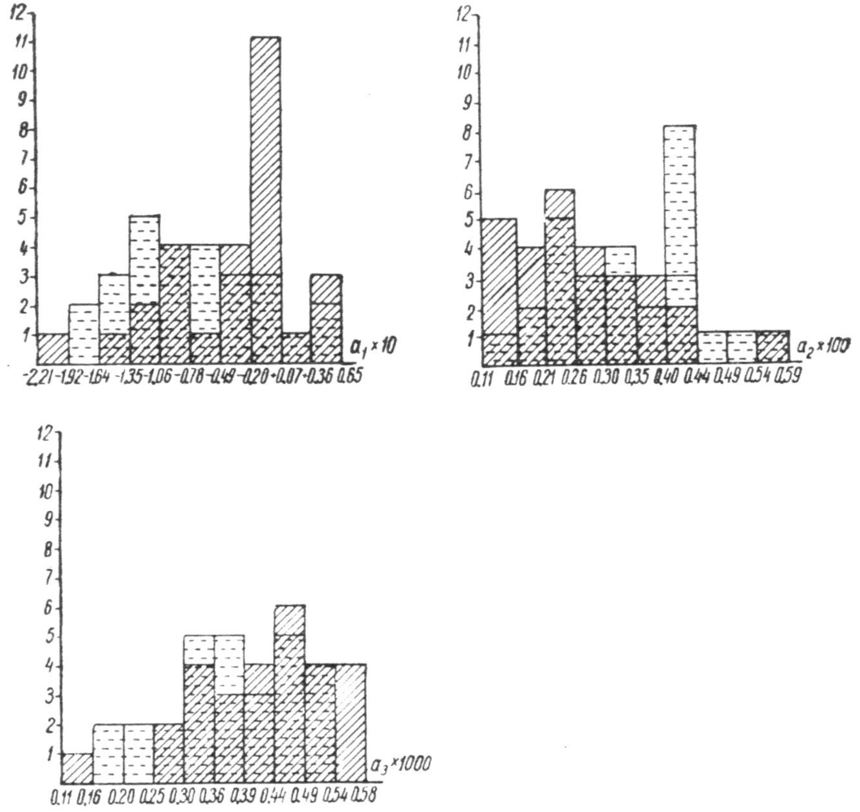

TABLE 33. Contingency for a_1

Source of sample	$a_1 \leqslant 0.4$	$a_1 > 0.4$	Σ
Bol'shoi Balkhan	$\dfrac{5^*}{14.5}$	$\dfrac{19}{13.5}$	28
Aman-Bulak	$\dfrac{20}{14.5}$	$\dfrac{8}{13.5}$	28
Σ	29	27	56

$$\chi^2 = 8.4.$$

* Here and henceforth the numerator gives the observed frequencies and the denominator the theoretical frequencies.

TABLE 34. Contingency for a_2

Source of sample	$a_2 \leqslant 0.3$	$a_2 > 0.3$	Σ
Bol'shoi Balkhan	$\dfrac{19}{14.5}$	$\dfrac{9}{13.5}$	28
Aman-Bulak	$\dfrac{10}{14.5}$	$\dfrac{18}{13.5}$	28
Σ	29	27	56

$$\chi^2 \approx 5.8.$$

TABLE 35. Contingency for a_3

Source of sample	$a_3 \leqslant -0.0132$	$a_3 > -0.0132$	Σ
Bol'shoi Balkhan	$\dfrac{11}{14.0}$	$\dfrac{17}{14.0}$	28
Aman-Bulak	$\dfrac{17}{14.0}$	$\dfrac{11}{14.0}$	28
Σ	28	28	56

$$\chi^2 \approx 2.6.$$

Fig. 72. Frequency distribution of polynomial coefficients of ρ_λ curves of samples from Middle Jurassic deposits of West Turkmenia. Dashes denote deposits from Bol'shoi Balkhan section, hatching denotes deposits from Aman-Bulak section.

TABLE 36. Table of Correlation Coefficients and Their Significances Between Photometric Parameters of Rocks and Their Content of Various Components from Two Middle Jurassic Sections in West Turkmenia

Photometric parameters	Aman-Bulak section				Bol'shoi Balkhan section						
	Fe_2O_3	FeO	mica	ore minerals (chromite + magnetite)	Fe_2O_3	FeO	hematite	quartz	rutile	carbonate content	clay fraction
$\overline{\varrho_1}$ 420–490	$\dfrac{-0.51}{2.8}$	+0.02	+0.20	+0.02	−0.25	$\dfrac{-0.41}{2.2}$	+0.34	$\dfrac{+0.36}{1.9}$	+0.30	−0.26	$\dfrac{-0.39}{2.1}$
$\overline{\varrho_2}$ 490–570	−0.30	+0.03	+0.20	+0.19	−0.20	−0.06	$\dfrac{+0.40}{2.1}$	$\dfrac{+0.38}{2.0}$	+0.30	$\dfrac{-0.38}{2.0}$	$\dfrac{-0.38}{2.0}$
$\overline{\varrho_3}$ 570–680	−0.21	+0.06	−0.12	0.00	+0.01	−0.27	$\dfrac{+0.43}{2.3}$	$\dfrac{+0.51}{2.9}$	+0.16	$\dfrac{-0.44}{2.4}$	−0.31
a_0	−0.14	+0.10	+0.00	+0.09	−0.04	−0.01	$\dfrac{+0.35}{1.9}$	0.19	+0.09	−0.24	−0.12
a_1	−0.31	−0.14	$\dfrac{+0.39}{2.1}$	$\dfrac{+0.42}{2.2}$	+0.02	+0.17	+0.02	0.03	$\dfrac{+0.43}{2.3}$	+0.16	−0.30
a_2	+0.19	+0.20	−0.24	$\dfrac{-0.37}{1.9}$	+0.03	+0.24	$\dfrac{+0.45}{2.4}$	0.18	−0.20	$\dfrac{-0.45}{2.4}$	+0.21
a_3	−0.10	−0.19	−0.15	+0.30	−0.19	+0.23	$\dfrac{-0.49}{2.6}$	−0.30	+0.12	$\dfrac{+0.52}{2.9}$	−0.13

Note: The numerator gives the values of the correlation coefficients and the denominator gives their significance evaluated by Fisher's method.

2. The values of the polynomial coefficients of the spectral luminance curves are affected as follows:

a) a_1 increases with increase in the amount of mica, ore minerals, and, in particular, rutile in the rock;

b) a_2 increases with increase in the quantity of hematite fragments in the rock and decreases with increase in the carbonate content of the rock;

c) a_3 increases with increase in the carbonate content of the rock and decreases with an increase in the quantity of hematite fragments in the rock.

Thus, the photometered samples of rocks of the same type from Middle Jurassic sections (from synchronous strata formed in different facies conditions) have different types of spectral luminance curves. The difference in the curves of externally similar samples of sandy-silt rocks was found to be significant by the χ^2 test. Hence, it follows that:

1. The parameters of the spectral luminance curves are typical for rocks of each facies environment. This indicates that the spectral luminance values of rock can be used for lithofacies mapping.

2. The parameters of the spectral luminance curves are correlated with the composition of the rock. The photometric parameters of Middle Jurassic sandy deposits are affected by their content of quartz, hematite, and soluble ferrous and ferric iron.

3. In aerial petrographic mapping the relationships between the clay content of the rock and its spectral luminance and between the carbonate content of the rock and its spectral luminance can be used.

Terrigenous Cretaceous Deposits of West Turkmenia

The spectral luminance of sandy-silt rocks of the Aptian, Albian, and Cenomanian stages in West Turkmenia was investigated by Vistelius (1958). In the investigation of the association between the photometric parameters of terrigenous Cretaceous deposits and their lithofacies characteristics we used some data from that work.

A section of Aptian and Albian terrigenous deposits representative of geosynclinal facies was investigated by Kravets. Below we give a stratigraphic description of a section of the Aptian and Albian (Fig. 73) on the southern slope of the Malyi Balkhan range [the Torengly section from the data of Kravets (1954)].

At the base of the section assigned to the Aptian stage lies a bed of gray compact calcareous sandstone containing brown phosphorite nodules. The fauna found in this sandstone consists of Deshayesites ex. gr. deshayesi Leym., D. ex. gr. weissi Neum. et Uhl., D. cf. consobrianus Orb., D. bodei Koen., and Sheloniceras ex. gr. albrechti Uhl.

The strata above the bed of sandstone with the ammonite fauna consist of black siltstones with the remains of oysters and sandy mudstones with bands of more compact sandstones containing oyster fragments.

The total thickness of the Lower Aptian is 226.3 m.

In the overlying Upper Aptian deposits the following strata can be distinguished (from bottom up):

1. Unit of dark gray clayey siltstones with beds of yellowish-gray thick-layered massive sandstone up to several meters thick. The interlayers of compact calcareous sandstone, which merge into sandy limestone, contain a fauna usually consisting of oyster fragments. These beds, which are most resistant to weathering, appear as ridges in the relief.

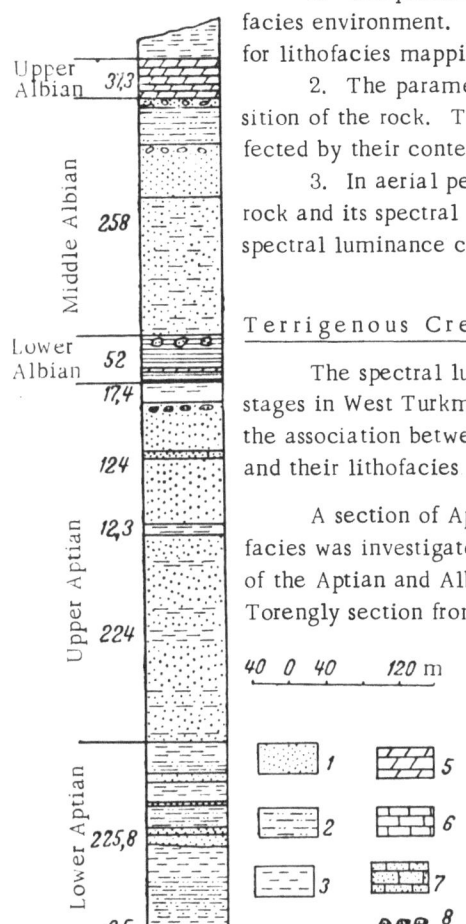

Fig. 73. Stratigraphic section of Lower Cretaceous (Aptian and Albian) deposits on southern slope of Malyi Balkhan range. Torengly section compiled by Kravets (1954). 1) Sandstones; 2) siltstones; 3) mudstones; 4) clays; 5) marls; 6) limestones; 7) sandy limestones; 8) concretions of sandy composition with fauna; 9) black clays at base of lower Albian.

103

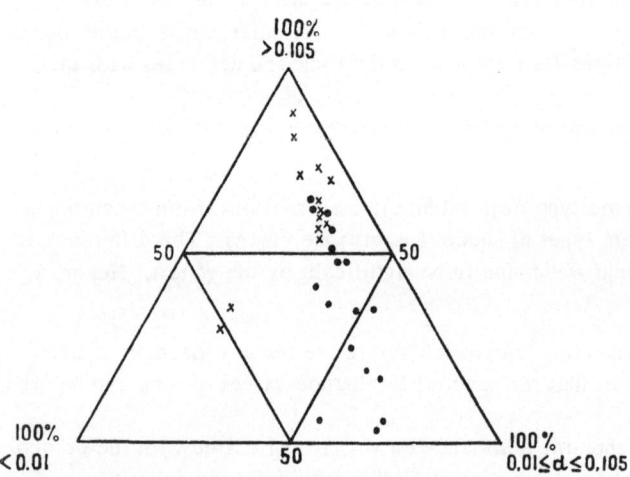

Fig. 74. Granulometric composition of samples from terrigenous Cretaceous deposits of West Turkmenia. Data of analyses of rocks from the Malyi Balkhan and Tuar sections are denoted by crosses and dots, respectively (from Vistelius, 1958).

2. Massive coarse-layered glauconitic sandstones with spherical concretions and compact calcareous sandstones with bands of clayey siltstones and clays. The concretions contain many casts of pelecypods. The solid calcareous sandstones contain trigonias, belemnites, and ammonites, represented by the dominant forms of the Upper Aptian: Parahoplites ex. gr. melchioris Anth., P. multicostatus Sinz., and Acanthoplites ex. gr. toblori Jacob.

The thickness of the Upper Aptian is 378 m. The total thickness of the terrigenous Aptian deposits is 604 m.

The Upper Aptian deposits give way gradually, with no traces of a break, to the Lower Albian deposits. Ammonites of the genus Acanthoplites, which appeared in the Upper Aptian, continued to exist in the bottom Lower Albian horizon, which consists of dark gray and black clays with a fine-layered texture and a thickness of 14 m. Above these clays lies a bed of light gray, slightly greenish marl containing belemnites; the top of the marl bed is compacted and covered with a brown crust of weathering. The marl is overlaid by a set of clays with concretions containing ammonites: Leymeriella tardefurcata Leym., Phylloceras sp., and Lytoceras sp. The ammonites Leymeriella tardefurcata Leym. are typical of the faunistic zone of the Lower Albian of the Caucasus, Mangyshlak Pen., and Western Europe.

The total thickness of the Lower Albian deposits is 52 m.

The Middle Albian deposits consist of lithologically uniform sandstone strata interbedded with siltstones and occasional bands of clay. Almost black clays containing concretions in which ammonites of the genus Leymeriella are found are exposed at the base of the Middle Albian deposits in the Torengly section. Above these clays lie greenish-gray, fine-grained sandstones, brownish on the surface, which give way to a thick set of weakly cemented fine-grained sandstone, alternating with sandy-clay layers. The sandstones have a typical greenish-yellow (tobacco) color and contain concretions of more compact sandstones. The concretions usually contain clusters of pelecypods and occasional ammonites. The total thickness of the Middle Albian deposits in the Torengly section reaches 260 m.

We did not investigate the Upper Albian deposits, which are represented in this section by bluish-gray marls with a fauna of dwarf ammonites.

For the spectrophotometric measurements we took samples from terrigenous deposits of the Aptian and Lower and Middle Albian – from beds of sandy and silty composition. From the whole section, which is about 1000 m thick, we collected and measured 18 samples.

Another section, synchronous with the above-described Torengly section of terrigenous Cretaceous deposits, was investigated in the region of Tuar well (see Fig. 69). From the Tuar section we took 14 samples, again from layers of sandy or silty composition. The thickness of this section does not exceed 220 m.

The description of the Aptian and Albian deposits begins east of the Irsary-Baba ridge and extends eastward to the Tuar well.

According to Vistelius (1958), the compared section from the regions of Tuar wells is as follows.

The Barremian red beds are exposed at the base of the Aptian deposits and are overlaid by (from bottom upward):

TABLE 37. Granulometric Composition and Carbonate Content of Samples from Compared Sections of Lower Cretaceous Deposits of West Turkmenia (data of Vistelius, 1958)

No. of sample	Carbonate content	Fractions, mm						
		0.210—0.149	0.149—0.105	0.105—0.074	0.074—0.053	0.053—0.01	<0.01	Totals
57	25.9	2.8	34.0	14.1	8.7	17.5	22.7	99.8
58	13.5	3.1	22.9	18.6	18.1	17.4	20.2	99.8
60	7.0	8.8	39.1	15.1	10.9	12.1	13.8	99.9
61	5.1	10.9	39.2	13.8	11.4	10.0	14.4	99.8
62	3.6	7.9	39.9	13.0	15.5	11.2	12.4	99.8
63	6.7	31.3	31.3	8.2	6.8	10.6	11.6	99.8
64	2.9	14.5	50.0	11.1	5.4	7.3	11.7	99.9
66	15.2	11.6	46.8	10.8	8.4	8.5	13.8	99.8
67	4.7	4.6	30.1	22.5	14.8	12.4	15.7	99.9
68	33.0	0.6	2.9	7.4	35.8	26.3	26.9	99.9
69	4.4	5.5	13.8	18.2	22.6	18.5	21.6	100.1
70	5.0	4.9	12.4	13.9	30.8	19.0	18.9	100.1
71	16.8	1.1	4.5	14.2	31.8	23.5	24.8	99.9
72	3.0	5.3	30.4	23.8	17.6	12.0	10.8	99.9
74	2.5	9.1	47.9	16.5	9.9	5.2	10.9	99.7
75	3.9	18.6	42.1	13.1	8.9	7.3	10.5	100.4
76	12.0	9.2	31.7	11.8	12.3	11.7	23.2	99.9
77	27.5	0.4	5.4	14.6	28.1	21.9	29.4	99.8
$\overline{X}$	10.7	8.3	29.1	14.5	16.5	14.0	17.4	
S	10.4	7.5	28.9	4.3	9.4	6.0	6.1	

Tuar

No. of sample	Carbonate content	Fractions, mm									
		>0,59	0,59—0,42	0,42—0,29	0,29—0,21	0,210—0,149	0,149—0,105	0,105—0,074	0,074—0,053	0,053—0,01	<0,01	Totals
112	23.5	—	—	—	0.1	5.8	29.8	5.9	1.7	10.5	45.8	99.8
113	1.8	—	—	—	0.1	1.2	58.3	16.4	7.6	3.5	12.2	99.4
114	4.8	—	—	—	0.1	9.4	62.1	9.1	4.8	2.9	11.2	99.7
116	35.1	—	—	—	0.9	60.5	26.3	2.9	2.0	1.9	5.2	99.9
117	5.1	—	—	0.1	0.3	1.7	29.2	10.2	5.4	3.3	50.6	99.9
119	1.7	—	—	—	0.2	6.9	62.6	14.1	7.0	3.2	5.5	99.5
121	1.2	—	—	0.2	0.5	20.5	54.9	11.3	5.1	2.4	4.7	99.7
122	1.5	—	—	0 1	0.3	16.7	59.3	11.6	5.1	2.5	4.3	99.8
123	11.7	—	—	0.1	0.2	54.6	19.0	11.2	5.1	3.2	6.1	99.7
124	2.2	—	0.1	0.3	4.4	47.9	27.8	5.1	3.6	2.3	8.1	99.6
125	40.0	—	—	0.1	0.1	4.6	54.3	13.5	7.4	6.5	13.2	99.8
127	31.6	—	—	—	0.3	17.1	44.0	10.9	7.9	7.9	11.6	99.8
128	38.9	—	—	—	0.1	11.1	51.9	12.3	7.5	5.8	11.2	99.9
129	44.5	—	—	0.1	0.1	8.2	47.0	12.3	9.9	2.6	14.6	99.8
$\overline{X}$	17.4				0.55	19.0	44.7	10.5	5.7	4.2	14.6	
S	17.3				1.5	21.2	15.2	6.5	2.3	2.4	14.9	

TABLE 38. Petrographic Composition of Rocks from Compared Sections of Terrigenous Cretaceous Deposits of West Turkmenia

No. of sample	Contents of fragments, %					
	potassic feldspar	quartz	plagio-clase	mica	rock fragments (including flints)	glauconite and authigenic chlorite
			Malyi Balkhan			
57	6.6	50.6	17.1	1.5	18.4	5.8
58	5.7	40.7	23.9	1.6	20.1 (8.0)	—
60	19.3	43.6	18.4	1.1	13.6 (3.1)	0.9
61	8.6	50.6	24.4	2.8	13.6	Rg
62	17.7	35.6	17.9	3.0	24.1	1.7
63	4.2	46.2	17.3	0.6	31.7	Rg
64	11.9	51.0	14.2	1.0	20.4	1.5
66	15.0	46.3	15.0	1.3	22.4	Rg
68	2.2	43.5	21.2	5.4	27.7	Rg
69	13.7	37.9	16.7	6.7	25.0	Rg
70	6.4	38.8	16.3	15.0	19.0	4.5
71	9.2	49.2	11.6	17.1	5.6 (7.3)	—
72	13.2	40.6	13.4	7.9	20.1 (4.8)	Rg
74	13.6	42.0	15.0	0.3	23.8	5.3
75	7.1	42.8	15.6	2.1	28.5	3.9
76	11.5	42.4	16.0	3.0	12.3 (5.5)	9.3
77	11.3	44.6	20.3	5.8	14.3 (3.7)	—
$\bar{X}$	10.4	43.9	17.3	4.5	20.0	
S	4.7	4.6	3.5	4.9	6.7	
			Tuar			
112	6.0	42.0	18.7	Rg	23.7 (9.6)	—
113	13.9	39.1	15.8	5.7	20.8	4.7
114	19.7	38.2	17.7	3.9	11.1 (9.4)	—
116	18.0	50.3	13.5	0.7	17.5	Rg
117	8.5	41.7	13.3	4.3	32.2	Rg
119	13.1	42.0	18.4	3.3	20.5	2.7
121	16.2	50.4	13.5	3.2	16.7	Rg
122	13.1	35.0	20.2	7.0	23.2	1.5
123	19.7	52.1	13.5	1.7	12.4	0.6
124	18.3	39.7	18.7	2.7	20.3	0.3
125	17.4	50.3	13.0	2.3	17.0	Rg
127	13.7	54.2	11.6	2.4	17.4	0.7
128	24.5	43.2	17.0	0.6	14.7	Rg
129	18.4	42.7	16.7	1.5	18.1	2.6
$\bar{X}$	15.8	44.4	15.8	2.8	19.0	
S	4.8	6.0	2.7	2.0	5.3	

TABLE 39. Mineralogical Composition of Heavy Fraction of Rocks from Two Compared Sections of Terrigenous Cretaceous Deposits of West Turkmenia (data of Vistelius, 1958)

Malyi Balkhan

No. of sample	Pyrite	Limonite	Hematite	Rutile	Anatase	Leucoxene	Ilmenite	Magnetite	Chromite	Zircon	Kyanite	Sillimanite	Staurolite	Amphibole	Monoclinic pyroxene	Epidote	Garnet	Tourmaline	Sphene	Monazite	Apatite	Barite
57	0.9	36.5	—	1.6	1.6	3.0	0.2	Rg	6.3	37.9			Rg	Rg	Rg	0.6	5.2	—	Rg	—	Rg	6.2
58	Rg	0.7	0.5	4.5	3.5	10.1	1.2	Rg	13.2	46.8	Rg		0.2	Rg	0.3	Rg	14.0	Rg	Rg	0.2	0.2	4.9
60		0.6	Rg	4.5	0.2	6.1	1.0	Rg	3.7	58.0			Rg	—	Rg	0.3	24.1	0.2	—	Rg	0.9	0.5
61		—	Rg	6.6	5.4	7.6	Rg	Rg	5.5	45.1	Rg		—	0.2	Rg	Rg	28.6	Rg			0.2	Rg
62		0.2	—	4.2	1.7	11.7	1.0	Rg	5.7	51.8	Rg		Rg	0.5	Rg	0.4	23.7	0.2			3.8	Rg
63		1.2	0.2	4.7	4.4	14.1	0.3	Rg	4.3	34.1			Rg	Rg	Rg	0.2	30.5	1.0			0.3	Rg
64		—	Rg	5.4	4.0	8.5	0.2	Rg	12.6	33.1			Rg	0.2	Rg	0.3	38.4	Rg			0.5	0.4
66	Rg	1.2	1.8	3.9	2.0	8.7	0.3	Rg	7.8	50.1			Rg	Rg	Rg	Rg	24.0	Rg	0.3		—	3.9
67		—	Rg	2.4	4.7	25.5	0.3	Rg	0.5	41.2		0.2	Rg	0.2	0.3	0.2	9.3	0.2	Rg	0.2	0.3	Rg
69		11.1	2.6	2.6	1.9	11.4	7.5	Rg	5.0	52.6			Rg	0.7	Rg	0.3	14.8	—	Rg		0.7	5.3
70		0.3	2.0	5.6	1.7	20.0	13.3	Rg	1.3	32.5	Rg		Rg	Rg	Rg	1.7	13.9	0.3	—	0.2	2.5	Rg
71		4.7	—	2.7	1.9	10.8	1.2	Rg	11.2	28.7			Rg	0.3	Rg	0.9	10.6	Rg	Rg	Rg	0.9	Rg
72		31.1	0.2	3.5	2.7	15.6	0.2	Rg	5.0	58.7	Rg		Rg	Rg	Rg	0.2	14.2	Rg	Rg		—	—
74		—	—	4.5	12.0	31.0	0.2	Rg	3.6	35.8			Rg	Rg	Rg	Rg	10.8	Rg			1.5	0.1
75		—	—	4.6	3.2	3.7	Rg	Rg	5.8	63.6		Rg	Rg	Rg	Rg	0.6	18.2	Rg	Rg			
76		44.4			0.2	9.6	—	Rg	0.6	27.8								10.0	0.3				
X̄				4.0	3.0	4.3	1.7		5.8	43.6						0.3	18.1				0.7	1.3
S				1.3	2.8	7.5	3.6		3.8	11.3						0.4	9.1					

Tuar

No. of sample	Pyrite	Limonite	Hematite	Rutile	Anatase	Leucoxene	Ilmenite	Magnetite	Chromite	Zircon	Kyanite	Sillimanite	Staurolite	Amphibole	Monoclinic pyroxene	Epidote	Garnet	Tourmaline	Sphene	Monazite	Apatite	Barite
112		Rg	0.7	1.1	0.2	—	44.4	2.3	0.4	8.5			0.4	—	—	0.7	4.7	Rg	0.5	Rg	0.3	0.8
113		Rg	1.7	0.7	0.1	—	74.5	5.6	0.3	8.7			0.3	Rg	—	0.3	6.0	Rg	Rg	Rg	Rg	Rg
114		Rg	Rg	1.0	0.8	—	70.1	3.8	0.6	10.5			0.8	—	—	0.8	8.7	Rg	0.6	Rg	Rg	Rg
116		—	1.1	Rg	0.3	—	66.4	7.9	0.3	11.0			0.5	—	—	0.5	9.2	0.3	0.3	Rg	0.1	Rg
117		Rg	Rg	0.3	0.1	—	50.3	4.7	Rg	6.4			0.2	Rg	—	0.2	6.8	—	Rg	Rg	0.1	—
119		Rg	—	3.8	0.7	—	54.4	3.0	Rg	7.7			2.7	—	—	3.2	9.8	0.6	0.6	Rg	0.4	0.2
121		—	0.9	1.3	0.4	—	66.1	3.5	Rg	7.8			1.2	—	—	0.3	7.5	—	0.6	Rg	Rg	1.5
122		Rg	0.8	0.9	0.6	—	67.0	3.9	Rg	6.3			0.9	Rg	—	Rg	11.2	Rg	0.3	Rg	Rg	0.6
123		Rg	0.5	1.2	0.4	—	72.7	5.0	Rg	8.8			Rg	Rg	—	Rg	8.1	—	0.6	Rg	Rg	Rg
124		0.3	1.3	1.4	0.1	—	67.6	5.0	Rg	6.7			1.6	—	—	Rg	11.8	—	3.4	Rg	0.1	0.1
125		Rg	2.7	0.5	0.2	—	67.4	Rg	Rg	7.5			0.3	Rg	Rg	0.3	2.7	Rg	Rg	Rg	Rg	Rg
127		0.5	3.4	0.6	0.1	—	62.7	3.0	—	5.4			1.0	Rg	Rg	0.2	15.1	Rg	Rg	Rg	—	7.0
128		1.8	5.6	1.4	0.1	—	61.8	4.0	Rg	3.4			1.2	Rg	Rg	0.6	6.1	Rg	0.2	Rg	Rg	7.5
129		Rg		0.7	0.1	—	61.8	6.0	0.2	9.2			2.2	Rg	Rg	0.5	8.6	Rg		Rg	Rg	14.2
X̄			1.3	1.1	0.3		63.4	3.9		7.7			1.0			0.5	8.3		0.5		0.8	2.3
S			2.6	0.91	0.06		8.5	1.97		4.0			0.77			0.77	3.0		0.91		0.76	4.2

107

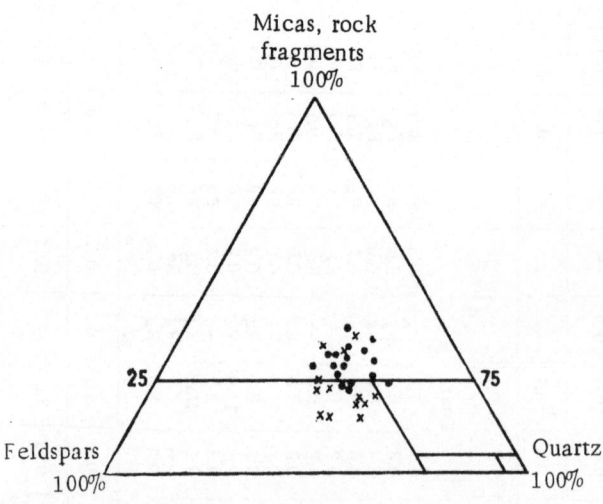

Fig. 75. Mineralogical composition of rocks from terrigenous Cretaceous deposits of West Turkmenia.
Crosses refer to samples from the Malyi Balkhan section; dots refer to samples from the Tuar section
(from Vistelius).

1. A set of fine-grained sandstones alternating with gray silty clays. The fauna found in this unit consists of ammonites of the genus Deshayesites. The thickness of the unit is 60 m.

2. Beds of brownish sandstones, containing lamel-libranchs (Trigonia archiacana Orb.), which characterize the top of the Lower and the bottom of the Upper Aptian. The unit is 70 m thick.

3. Beds of sandstones with a fauna of Upper Aptian ammonites: Parahoplites melchioris Anth., Acathoplites (Colombiceras) tobleri Jacob., etc. The thickness of these beds is about 50 m.

The total thickness of the Aptian deposits in the Tuar section is 180 m.

The deposits of the Albian stage merge gradually into the Upper Aptian sandstones and it is difficult to trace the boundary between them. They also consist mainly of fine-grained yellowish-gray, sometimes red-dish or brownish sandstones. The cement of the sand-stones is so weak that in the zone of oxidation they are transformed into loose sands. A characteristic feature of these beds is the presence of large calcareous-sandy concretions of spherical or rounded cylindrical shape. The fauna found in these concretions consists of ammonites (Acanthoplites nolani Seun.) and lamellibranchs (Aucellina caucasica Buch., Thetironia sp.), gastropods, and, occasionally, echinoids.

The sandstones alternate with clay beds, which occur only in the lower substage. The thickness of the Lower Albian in the Tuar section is 120-140 m.

The Middle Albian deposits consit of fine-grained pinkish sandstones containing Inoceramus. The thick-ness of the Middle Albian does not exceed 100 m. Upper Albian deposits did not come into the investigated section.

The total thickness of the terrigenous Aptian and Albian deposits in the Tuar section is 220 m. The section was characterized by 14 samples taken at more or less equal intervals.

The results of investigations of the rock samples from terrigenous Cretaceous deposits of West Turkmenia are given below. Parallel characterizations of the samples from the two compared sections (the Malyi Balkhan section representing the geosynclinal facies and the Tuar section representing the platform facies) are given.

Petrographic-Mineralogical Characterization of Terrigenous Cretaceous Deposits from the Tuar and Malyi Balkhan Sections

Analyses of the samples taken from the sections showed that the carbonate content of sandy-silt rocks from the Tuar section is much higher on the average than that of similar deposits from the Malyi Balkhan section. The granulometric composition of the samples from these sections (Table 37) is also different. The samples from the Tuar section consist mainly of silty or clayey-silty sands, whereas samples from the Malyi Balkhan section con-sist mainly of sandstones (Fig. 74).

In petrographic composition (Table 38) the investigated samples are graywackes or arkoses. The rocks from the Tuar section are more like arkoses, while those from the Malyi Balkhan section are closer to graywackes (Fig. 75).

The content of quartz fragments in the rock samples from the compared sections is approximately the same. There is no significant difference in the total content of feldspars either. The amount of lime feldspars is higher

TABLE 40. Soluble-Iron Content of Samples from Terrigenous Cretaceous Deposits of West Turkmenia (from data of Vistelius, 1958)

No. of sample	Hygroscopy	Total iron content	FeO	Fe_2O_3
		Malyi Balkhan		
57	2.34	1.88	1.01	0.76
58	2.40	2.64	1.54	0.83
60	0.77	2.47	1.26	1.07
61	1.04	2.58	1.19	1.26
62	0.69	2.82	1.50	1.15
63	0.96	2.08	1.09	0.87
64	0.67	1.69	0.60	1.02
66	0.63	1.90	0.53	1.31
67	0.82	2.21	0.62	1.52
68	0.58	1.92	0.96	0.85
69	1.25	2.92	0.65	2.20
70	1.09	2.64	0.91	1.62
71	1.34	2.40	1.03	1.25
72	1.06	2.53	0.46	2.02
74	0.49	1.93	0.75	1.09
75	0.93	1.92	0.69	1.16
76	2.10	3.01	0.86	2.05
77	0.68	2.40	1.24	1.02
$\bar{X}$	1.10	2.33	0.94	1.28
S	0.59	0.40	0.32	0.44
		Tuar		
112	3.31	1.50	0.16	1.32
113	0.98	2.43	0.34	2.05
114	1.43	2.41	0.41	1.96
116	0.59	1.55	0.05	1.55
117	1.52	4.39	0.00	4.39
119	0.68	1.72	0.32	1.40
121	0.57	1.71	0.27	1.41
122	0.59	1.61	0.24	1.34
123	0.59	1.35	0.31	1.00
124	0.77	2.38	0.20	2.16
125	0.63	1.59	0.05	1.59
127	0.55	1.40	0.03	1.37
128	0.92	1.48	0.08	1.39
129	0.84	2.29	0.23	2.03
$\bar{X}$	1.00	1.99	0.19	1.78
S	0.73	0.80	0.13	0.82

TABLE 41. Values of Spectral Luminance Factors of Rocks from Terrigenous Cretaceous Deposits of West Turkmenia (from Vistelius, 1958)

No. of sample	Spectral luminance factors (%) for each wavelength										
	450	467	490	510	534	570	594	604	638	672	699
	Malyi Balkhan										
57	22.3	23.3	25.5	27.5	29.3	31.6	31.0	32.5	31.8	30.0	28.0
58	17.6	18.6	20.8	22.5	23.6	26.4	27.4	27.5	26.5	26.0	25.0
60	19.5	20.5	23.0	26.0	28.0	31.8	32.5	34.3	31.6	30.6	29.0
61	16.6	17.2	20.0	22.6	25.3	29.5	29.6	30.8	29.6	28.5	28.4
62	19.8	19.5	21.3	24.5	27.1	31.5	32.5	32.8	32.5	32.0	30.5
63	20.0	20.2	23.5	24.5	27.2	30.3	30.5	31.9	30.3	28.3	31.0
64	16.1	16.0	18.6	20.6	22.6	26.1	26.5	27.3	25.8	27.0	25.0
66	18.5	18.1	20.1	22.1	24.0	27.0	26.8	29.9	28.6	29.0	26.8
68	26.2	26.0	27.6	30.3	31.6	35.6	37.0	37.0	35.3	36.8	34.3
69	14.4	15.0	15.5	17.8	18.0	22.2	22.2	22.5	22.5	22.0	20.8
70	17.5	17.8	20.0	23.3	25.6	30.3	30.3	31.0	31.3	32.0	31.4
71	20.0	20.0	21.0	25.0	25.6	30.0	30.0	30.0	30.0	31.6	30.0
72	14.6	15.0	18.0	21.5	24.3	29.0	31.6	31.2	30.6	30.0	27.6
74	17.8	19.3	21.6	23.4	27.0	31.0	31.0	31.3	30.3	30.0	30.0
75	18.3	19.3	21.2	23.3	25.3	30.3	29.8	37.1	30.0	28.3	30.0
76	14.0	15.6	17.8	18.8	20.8	22.8	22.1	31.1	20.7	20.0	19.3
77	19.2	20.0	22.8	23.6	25.0	27.4	27.3	27.2	26.6	26.0	28.3
$\bar{X}$	18.4	18.9	21.1	23.4	25.3	29.0	29.3	30.9	29.1	28.7	28.0
S	3.01	2.87	2.91	2.99	3.16	3.37	3.69	3.58	3.66	3.90	3.77
	Tuar										
112	18.0	13.0	21.0	24.2	27.5	35.0	36.1	36.0	39.0	40.0	39.3
113	12.6	13.2	16.4	18.6	23.7	31.0	34.1	34.1	35.0	34.5	36.1
114	12.6	13.0	16.1	18.6	22.6	30.2	31.6	31.0	32.6	31.5	34.0
117	17.0	18.0	20.5	21.8	21.8	29.7	30.3	30.3	30.8	31.7	30.5
119	15.0	15.7	17.8	20.0	24.1	29.1	31.3	30.2	31.6	30.0	31.0
121	14.9	15.0	18.0	20.3	23.8	28.9	31.8	33.3	32.0	32.7	32.2
122	14.7	15.2	17.7	19.3	23.2	28.3	30.6	30.2	31.0	31.1	31.0
123	15.5	16.5	18.0	19.3	21.4	25.6	27.3	27.8	25.6	28.2	28.5
124	12.7	13.4	16.0	18.0	21.0	27.0	28.7	28.0	29.6	30.0	29.6
125	13.3	14.0	17.6	19.9	25.2	28.8	27.7	27.7	28.3	27.0	14.6
127	13.9	14.9	16.5	17.2	20.2	23.9	26.2	26.0	26.6	26.7	26.2
128	14.6	15.3	17.0	19.0	21.0	25.0	28.2	28.7	29.2	30.0	29.0
$\bar{X}$	14.6	14.8	17.7	19.7	23.0	28.5	30.3	30.3	30.9	31.1	30.2
S	1.71	1.54	1.60	1.85	2.08	2.98	2.90	2.95	3.63	3.59	6.04

TABLE 42. Values of Polynomial Coefficients of ρ_λ Curves of Samples from Terrigenous Cretaceous Deposits of West Turkmenia (from data of Vistelius, 1958)

No. of sample	a_0	a_1	a_2	a_3
		Malyi Balkhan		
57	0.1937	0.0261	—0.0012	—0.00002
58	0.1573	0.0185	0.0112	—0.00006
60	0.1076	0.0475	—0.0014	—0.00011
61	0.1429	0.0201	0.0007	—0.00010
62	0.1784	0.0099	0.0024	—0.00018
63	0.0038	0.0276	—0.0012	—0.00001
64	0.1292	0.0247	—0.0007	—0.00003
66	0.1721	0.0083	0.0017	—0.00010
68	0.2307	0.0237	—0.0008	—0.00015
69	0.1316	0.0057	0.0018	—0.00011
70	0.1538	0.0177	0.0010	—0.00010
71	0.1735	0.0166	0.0005	—0.00008
72	0.1193	0.0147	0.0027	—0.00022
74	0.1516	0.0251	—0.0002	—0.00006
75	0.1542	0.0222	—	—0.00006
76	0.1593	—0.0001	0.0020	—0.00015
77	0.1820	0.0187	—0.0014	0.00004
$\overline{X}$	0.1495	0.0192	0.0010	—0.000088
S	0.0477	0.0106	0.0030	0.000065
		Tuar		
112	0.1519	—0.0173	0.0023	—0.00017
113	0.1056	0.0155	0.0033	—0.00024
114	0.0887	0.0255	0.0014	—0.00016
116	0.1140	0.0199	0.0004	—0.00006
117	0.1460	0.0051	0.0028	—0.00016
119	0.1207	0.0018	0.0022	—0.00024
121	0.1304	0.0114	0.0026	—0.00010
122	0.1348	0.0111	0.0024	—0.00018
123	0.1150	0.0277	—0.0016	0.00004
124	0.1043	0.0176	0.0015	—0.00010
125	0.1145	0.0146	0.0035	—0.00030
127	0.1366	0.0004	—0.0049	—0.00019
123	0.1408	0.0025	0.0033	—0.00020
129	0.1809	—0.0140	0.0035	—0.00019
$\overline{X}$	0.0470	0.0128	0.0022	—0.00008
S	0.0235	0.0133	0.0023	0.00008

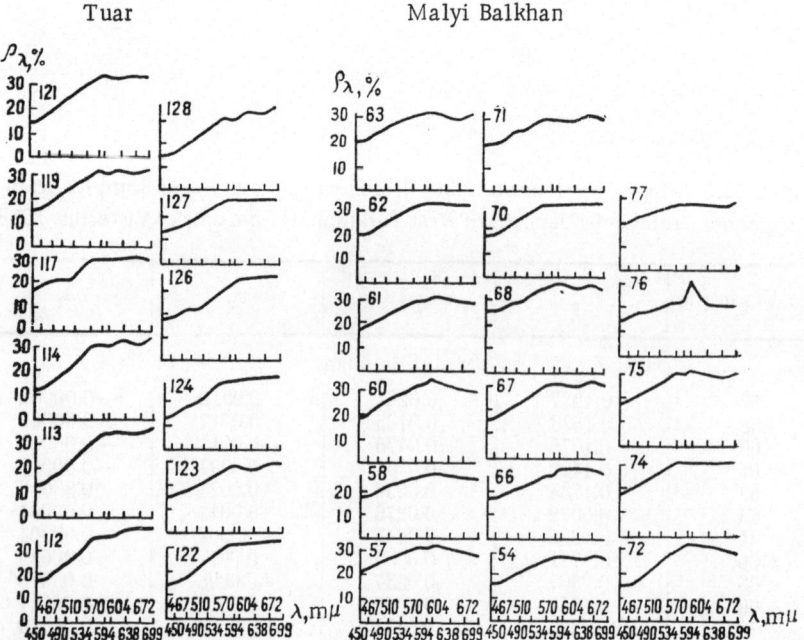

Fig. 76. Nature of ρ_λ curves of samples from compared sections of terrigenous Cretaceous deposits of West Turkmenia (from Vistelius, 1958).

than that of predominantly potassic feldspars in the Malyi Balkhan section, and the content of predominantly potassic feldspars is slightly higher than that of lime feldspars in the Tuar section. The sandstones from the Malyi Balkhan section are appreciably richer in mica fragments and glauconite.

An analysis of Table 39 shows that the heavy fraction of samples from the Malyi Balkhan section is rich in limonite, chromite, zircon and leucoxene, whereas the heavy fraction of samples from the Tuar section is rich in ilmenite and magnetite. Limonite occurs mainly as rare grains.

The total soluble iron content of the samples from the compared sections is different (Table 40). In samples from Malyi Balkhan section the ferrous iron content is greater than the ferric iron content, while in the samples from the Tuar section the reverse is true.

The results of photometry of the samples from terrigenous Cretaceous deposits of West Turkmenia are given in Table 41 in the form of the values of the spectral luminance factors of the analyzed samples and in Table 42, which gives the values of the polynomial coefficients of the spectral luminance curves of the same samples. The appearance of typical ρ_λ curves of the samples from the compared sections is shown in Fig. 76.

The values of ρ_λ show that samples from the Malyi Balkhan section reflect more strongly in the shortwave part of the spectrum (in the wavelength range 490-510 mμ, corresponding to the green part of the spectrum), whereas samples from the Tuar section reflect more strongly in the long-wave part of the spectrum (in the wavelength range 670-699 mμ).

Figure 77 shows the frequency distribution of values of the polynomial coefficients of the ρ_λ curves of rocks from the compared sections. The nature of these distributions indicates that the distributions of a_2 and a_3 are made up of two separate distributions; i.e., they can be used for distinguishing the lithofacies types of the investigated deposits.

Tests of the difference in the values of the polynomial coefficients by the χ^2 method are shown in Table 43, which indicates that these differences are significant.

The relationship between the composition of the rocks and the values of the polynomial coefficients of the ρ_λ curves of samples from terrigenous Cretaceous deposits of West Turkmenia was determined by calculation

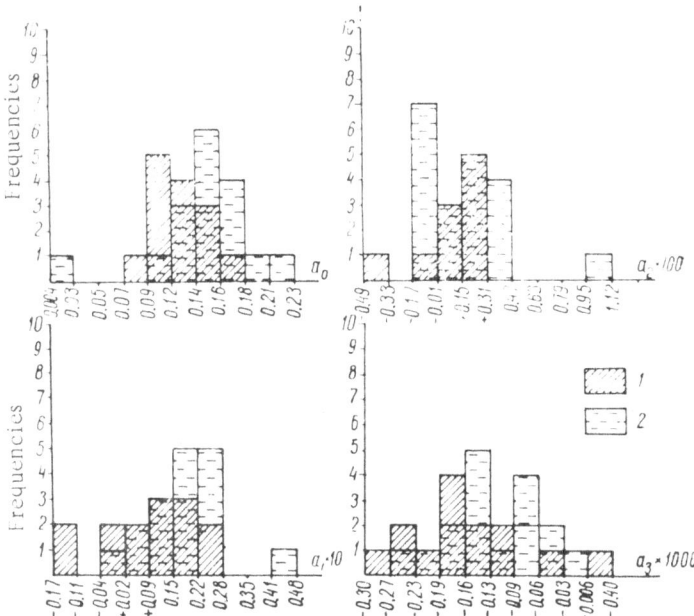

Fig. 77. Frequency distribution of polynomial coefficients of ρ_λ curves of terrigenous Cretaceous deposits of West Turkmenia. 1) Samples from Malyi Balkhan section; 2) samples from Tuar section.

of the correlation coefficients. The obtained values of the correlation coefficients showed that the value of a_3 depends mainly on the amount of chromite and quartz in the rock. No other associations could be detected in view of the small number of observations.

Thus the results of laboratory measurements of the spectral luminance factors of samples from synchronous terrigenous Cretaceous strata of West Turkmenia, formed in conditions close to typical geosynclinal and typical platform conditions, revealed a difference in the ρ_λ values of sandy-silt rocks which are externally similar to one another. The difference revealed in the evaluation of the spectral luminance curves of these deposits permits their separation into lithofacies types.

Tertiary Deposits of West Turkmenia

The third material on which we tested the possibility of mapping lithofacies types of terrigenous deposits from their spectral luminance was West Turkmenian Neogene deposits in regions with different conditions of sedimentation.

By the beginning of the Neogene epoch the paleogeographic environment within the considered territory had altered appreciably. The previously prevailing conditions of marine sedimentation had given way mainly to conditions of continental sedimentation. Part of the region belonging to the platform was transformed into dry land, and in the transitional zone there appeared the Kubadag – Bol'shoi Balkhan mountain system. The region of the Alpine geosyncline continued to undergo considerable subsidence, particularly in its western part.

We investigated facies complexes of Neogene deposits in different regions: on Krasnovodsk Peninsula, on Cheleken Peninsula, and in separate structures south of Nebit-Dag (Syrtlanli, Boya-Dag, Monzhukly, etc.).

To tackle our problem, i.e., to compare synchronous deposits which had accumulated in conditions close to geosynclinal and typical platform conditions, we can use the data of an investigation of the Syrtlanli section, which represents the geosynclinal facies, and sections of the red beds from the region of the southern scarp of Krasnovodsk Peninsula (Belek and Kara-Tengir sections), which represent platform facies. The location of these sections is shown in Fig. 67.

Syrtlanli Section

The section of red beds exposed in the core of the anticlinal fold in the Syrtlanli region consists mainly of very coarse or coarse, poorly sorted sands, with occasional sandstones containing inclusions of clayey pebbles (balls) interbedded with brick-red clays. A stratigraphic description of the section made in 1956 is shown in Fig. 78, which also indicates the source of the samples for the photometric measurements. The total thickness of the section is 47 m. The uniformity of the deposits, the constancy of their composition along the strike, and their great thickness characterize the section of red beds in the Syrtlanli region and are characteristic of geosynclinal conditions of sedimentation.

The data of mineralogical analyses of the heavy fraction of samples from the Syrtlanli section are given in Table 44. The data of the photometric measurements are given in the form of spectral luminance factors in

TABLE 43. Contingencies Between Polynomial Coefficients a_1, a_2 and a_3 and Region of Accumulation of Investigated Rock from Terrigenous Cretaceous Deposits of West Turkmenia

Region	$a_{1_0} < 0.0175$	$a_{1_0} > 0.0175$	Σ
Malyi Balkhan	$\dfrac{11}{8.226}$	$\dfrac{6}{8.774}$	17
Tuar	$\dfrac{4}{6,774}$	$\dfrac{10}{7.226}$	14
Σ	15	16	31

$$\chi^2 = 4.01; \; P(\chi^2) \geqslant 0.0163.$$

Region	$a_{2_0} \leqslant 0.0011$	$a_{2_0} > 0.0011$	Σ
Malyi Balkhan	$\dfrac{10}{8.226}$	$\dfrac{7}{8.774}$	17
Tuar	$\dfrac{5}{6.774}$	$\dfrac{9}{7.226}$	14
Σ	15	16	31

$$\chi^2 = 1.64; \; P(\chi^2) \geqslant 0.20.$$

Region	$a_{3_0} \leqslant 0.000010$	$a_{3_0} > 0.000010$	Σ
Malyi Balkhan	$\dfrac{6}{8.774}$	$\dfrac{11}{8.226}$	17
Tuar	$\dfrac{10}{7.226}$	$\dfrac{4}{6.724}$	14
Σ	16	15	31

$$\chi^2 = 4.01; \; P(\chi^2) \geqslant 0.0163.$$

Table 45 and in the form of polynomial coefficients of the ρ_λ curves in Tables 46 and 47.

Sections of Red Beds in Belek and Kara-Tengir Regions on Krasnovodsk Peninsula

For photometric measurements we investigated two sections which supplement one another (Belek and Kara-Tengir), situated on the southern slope of the Kyuryanyn-Kyure scarp. The description of the section was made from strippings and from separate prospect holes, which were instrumentally linked with the relief of the locality. The stratigraphic columns of these sections are shown in Fig. 79.

The red bed outcrops on the Kyuryanyn-Kyure scarp consist mainly of conglomerates and gravel polymictic sandstones with bands of brown clays and loams.

In view of the fact that the material of the sediments consists mainly of large fragments, we took sieved fractions with a particle size of not more than 5 mm for the photometric measurements. The heavy residues for petrographic analysis were obtained from the sieved fractions by washing and were investigated in the usual way. The petrographic composition of the pebbles and boulders was determined visually. For the quantitative

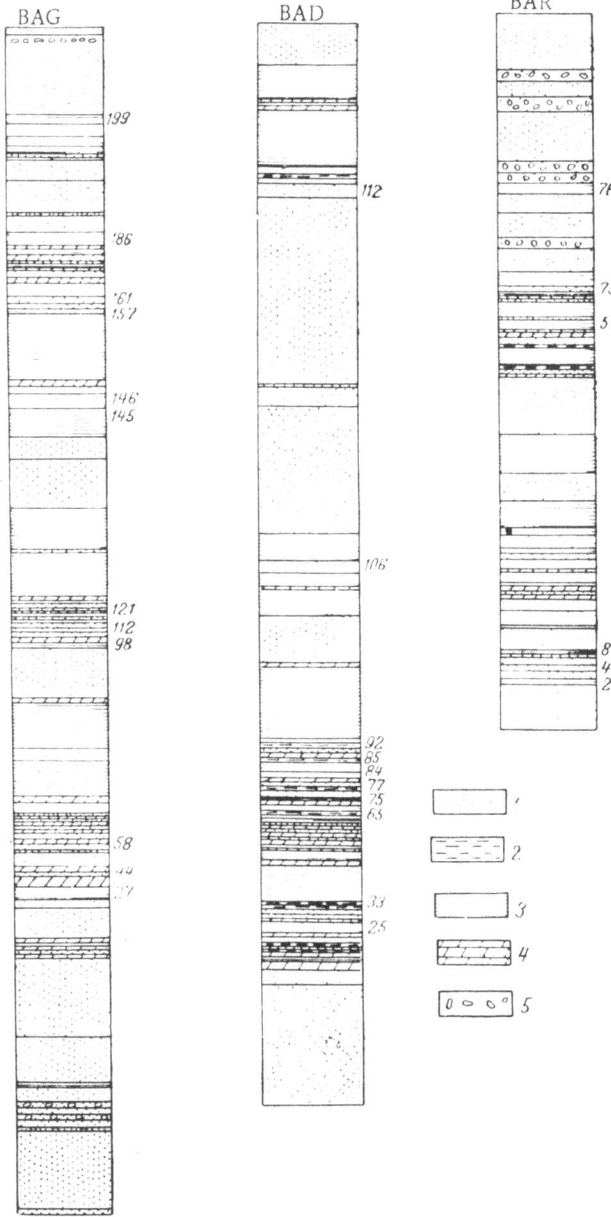

Fig. 78. Stratigraphic sections of upper part of red beds in Syrtlanli region. 1) Sands; 2) siltstones; 3) clays; 4) siltstones alternating with clays; 5) clay balls in sand. The figures beside the columns give the sample numbers.

determination of the composition of the pebbles we took a sample pailful of conglomerate, passed it through sieves, and used all the pebbles larger than 5 mm in the determination of the composition of the fragmentary material of the sample.

A description of the investigated sections in the region of Belek wells and north of the Kara-Tengir station is given below.

Belek Section. The description of the section begins in the immediate vicinity of the easternmost well, where limestones and calcareous sandstones with an abundant fauna characteristic of the Konkian horizon are exposed on the southern scarp of Kyuryanyn-Kyure (Vistelius and Korobkov, 1953). From this section we obtained the following samples:

Sample B-3 – from a bed of conglomerate with large (up to 25 cm), well-rounded limestone pebbles and almost unrounded granitoid pebbles. The cement of the conglomerates is clayey-sandy. The sample for photometric measurements was taken from the sieved part of the sandy cement. The heavy fraction obtained from this sample consists mainly of magnetite, ilmenite, hematite, and limonite and contains a fair amount of zircon. The thickness of the conglomerate bed was 2.3 m.

Sample B-4 – from a bed of conglomerate lying on a thick bed of clays. The conglomerate pebbles from this layer are small, not more than 5-6 cm in diameter. The pebbles are composed mainly of limestones and porphyritic granites. The minerals in the heavy fraction washed from the clayey cement of these conglomerates consist mainly of magnetite, limonite. and epidote.

Sample B-5 – from a bed approaching clay in composition and containing sparse pebbles of limestones, granites, and sandstones. The heavy residue washed from this clay consists of magnetite, ilmenite, and limonite, and has a characteristically high content of barite (18.3%) in the form of large transparent crystals, sometimes with a zonal structure. Many large grains of sphene are present.

Sample B-6 – from a bed of clayey conglomerates, highly gypsiferous, and containing inclusions of small pebbles, mainly of granite composition. The pebbles are not more than 8 cm in diameter. The heavy fraction from the cement of this conglomerate is rich in magnetite and zircon.

Sample B-7 – from a bed of fine-pebble conglomerates with compact calcareous cement. The individual pebbles reach 5-6 cm in diameter. The heavy fraction contains a large amount of ore minerals and zircon, and the presence of barite is characteristic.

Sample B-9 – from a bed of conglomerate lying on a layer of brown clays. The largest pebbles are granitoid (forming up to 80% of the total number of pebbles in the sample). Calcareous pebbles constitute about 2%. The heavy fraction consists of large (1.7 mm) grains of magnetite, which is often found in aggregates with quartz.

TABLE 44. Mineralogical Composition of Heavy Fraction from Red-Bed Deposits in Syrtlanli, Belek, and Kara-Tengir Sections (West Turkmenia)

Syrtlanli

No. of heavy residue	Corundum	Martite	Hematite	Limonite	Ilmenite	Rutile	Anatase	Leucoxene	Chromite	Magnetite	Kyanite	Staurolite	Zircon	Amphibole group	Orthorhombic pyroxene group	Monoclinic pyroxene group	Garnet group	Epidote group	Tourmaline	Chlorite	Sphene	Apatite	Carbonate group	Barite	Mica	Rock fragments
BAG-2			7.6		16.8	0.2		0.1	0.4	3.6	0.2	1.1	2.3	14.5	0.2	0.5	30.2	21.6	0.4	Rg			0.2			0.1
BAG-5	0.1		3.5	5.0	10.4	1.1		1.8		2.1	0.3	0.2	3.5	17.4	0.3	0.4	41.5	16.5	Rg	0.5	0.1	0.4				
BAG-24			3.9	9.6	9.7	0.7		2.4		2.9	0.3	Rg	4.3	18.5		1.0	29.2	20.0	1.0	Rg		0.6		7.5		
BAG-44			3.8	3.6	5.2	2.5	0.2	23.9		Rg	0.5		12.6	5.3	0.7	0.9	17.7	5.7	0.4	1.7		0.8		2.7	1.0	3.3
BAG-89			2.1		2.8	2.1	0.4	1.3		0.3	0.2		18.0	2.6	1.4	1.8	18.1	5.7	Rg	Rg		0.7				35.6
BAG-94		0.8	5.4	2.7	10.0	1.4		1.1	0.3				0.5	33.4		1.0	17.0	16.0				Rg	0.7			6.7
BAG-141			16.2		30.2	1.7		1.2		5.7	0.1	1.2	5.1	2.8	0.9	1.0	26.2	6.0	0.4	Rg	0.1	0.2	2.0			
BAG-155			13.7	6.6	20.2	0.9	Rg	0.9		1.2			0.4	14.9		1.0	10.1	26.2	Rg		0.4		Rg	0.9		
BAG-191			17.5		20.2	0.9		1.2		5.9	0.1	0.2	6.3	3.8		1.3	30.6	11.2	0.3		0.9	0.3	2.5	0.4		
BAG-191			17.3		22.9	1.3	0.02	1.0		5.9			6.9	6.9	2.4	2.4	21.6	9.9	0.9	0.6	0.1	0.2		Rg		
BAG-200			8.1		16.0	2.1		1.4		2.8	0.5	1.6	2.0	11.9	1.1	2.2	31.3	15.6		0.2		0.1	5.2			
BAG-200		0.5	9.3		14.3	0.9	0.1	1.0		2.1	0.3	0.4	1.8	5.7	1.0	2.2	38.1	10.2	0.2		0.1	0.2	Rg			
BAD-36			12.7	6.3	11.0	1.2		0.2	0.5	2.1	0.1	0.7	3.1	7.8	2.4	1.0	35.4	18.6	0.3		0.1	0.5				
BAD-84			18.8	5.4	36.5	1.0		2.1	0.8	1.5			2.1	3.1		3.1	17.7	3.1			1.1					
BAD-98			9.2		19.2	0.8	0.1	1.5	0.5	5.4		Rg	Rg	10.0		1.4	29.2	17.7				0.8		0.1	0.8	
BAD-109			14.1	4.1	16.2	0.6		0.4		1.3	0.9	0.5	4.0	7.8	0.1	1.7	27.0	23.7	1.2	Rg	0.1	0.2			0.8	
BAD-111			5.3		5.7	0.2		2.9		2.0	0.4	1.7	2.0	34.8	0.2	3.6	12.1	28.9	1.2			0.6			0.8	
BAD-117			15.8	4.9		2.4				3.5	1.8	0.6	3.7	24.4		2.8	23.1	17.1	2.2		0.2				0.5	
BAD-132			9.9	0.1	11.0	4.7	0.1	2.9		1.5	0.1	2.1	0.4	16.5	1.1	1.7	20.9	22.0	0.4		1.9	0.6				
BAE-1			9.3		5.5	0.8		0.9		0.5	0.3		1.4	6.8	0.5	1.6	31.8	28.1	0.6			0.4				6.7
BAE-41	2.2		8.7		19.8	0.5		1.6					1.1	5.5			33.4	23.5							0.5	
BAE-42			21.3		25.9	1.8		2.9		3.5	1.8	0.6	0.6	6.4			20.6	13.4		1.1				0.6		
$\bar{X}$			10.6		15.0	1.4		2.4		2.6			3.7	11.8		1.6	25.6	16.4								
s			5.6		9.1	1.0		4.9		1.9			4.3	9.3		0.9	8.4	7.5								

116

TABLE 44 (cont'd). Mineralogical Composition of Heavy Fraction from Red-Bed Deposits in Syrtlanli, Belek, and Kara-Tengir Sections (West Turkmenia)

No. of heavy residue	Grain size, mm	Limonite	Hematite	Martite	Rutile	Nigrine	Anatase	Leucoxene	Ilmenite	Magnetite	Chromite	Zircon	Kyanite	Staurolite	Hornblende	Epidote	Garnet	Tourmaline	Biotite	Chlorite	Sphene	Tantaloniobate	Barite	Celestite	Carbonates	Rock fragments
Belek																										
B-3	0.05—0.80	13.7	5.5	—	Rg	Rg	—	0.8	21.8	53.0	—	1.6	—	0.1	—	2.5	1.0	—	—	—	Rg	—	—	0.4	0.9	—
B-4d	0.05—1.5	37.1	—	—	Rg	Rg	—	4.5	—	44.6	—	1.0	—	—	—	8.7	2.6	—	—	—	0.2	—	—	—	0.4	—
B-5	to 1mm	15.3	5.6	2.2	—	Rg	0.1	0.7	13.0	33.1	—	9.1	—	—	—	1.9	—	—	—	—	0.5	—	18.3	0.3	0.1	0.2
B-6b	0.01—0.5	6.0	7.0	—	Rg	Rg	—	0.5	15.0	64.2	—	4.1	—	—	—	1.3	0.1	—	—	—	0.7	—	—	—	—	0.7
B-7	—	13.1	8.5	—	Rg	Rg	—	—	24.6	48.6	—	1.6	—	—	—	1.5	—	—	—	—	0.2	—	1.2	—	—	0.6
B-9	0.05—1.7	13.0	11.0	0.1	—	—	—	0.6	11.6	26.7	—	1.6	—	—	Rg	2.5	0.2	—	—	—	0.4	—	31.5	0.4	0.5	—
B-14	—	13.6	10.3	2.8	Rg	Rg	—	—	1.0	69.0	—	1.4	—	—	—	3.2	—	—	—	—	0.4	—	—	0.5	0.4	—
B-15	—	4.9	8.4	—	Rg	Rg	—	Rg	20.6	57.7	—	1.4	—	—	—	2.8	0.8	—	—	—	0.8	—	0.2	0.3	—	—
B-17	0.1—1.5	3.8	8.9	—	0.1	Rg	—	—	15.8	65.4	0.3	1.5	—	—	—	1.9	0.4	—	—	—	1.4	—	0.3	0.4	—	—
B-19	—	4.1	2.9	2.9	Rg	—	—	Rg	24.0	63.8	—	1.7	0.1	—	—	1.5	0.3	—	—	—	0.9	—	—	0.2	—	—
B-21	—	2.0	12.8	—	—	Rg	Rg	0.2	9.4	70.8	—	2.1	—	—	—	0.5	—	—	—	—	1.5	—	0.2	0.1	—	—
B-23	to 1 mm	4.3	9.2	—	Rg	Rg	—	—	13.0	66.1	—	0.7	Rg	—	—	1.6	—	—	—	—	1.8	—	—	0.1	—	—
B-24	—	3.1	7.1	2.4	—	Rg	—	Rg	16.4	69.0	—	1.6	Rg	—	—	0.9	0.3	—	—	—	1.8	—	0.4	1.6	0.2	—
B-26	—	2.8	1.9	—	0.2	—	Rg	0.1	10.2	75.4	0.1	0.8	—	—	—	2.1	Rg	—	Rg	Rg	1.9	—	0.1	0.5	—	1.4
B-27	—	4.1	10.9	—	0.2	—	Rg	0.2	6.0	71.5	—	1.5	—	—	0.2	2.9	0.2	—	Rg	0.1	0.8	—	0.1	—	—	—
B-28	—	6.2	5.7	1.4	0.1	—	—	—	7.5	72.7	—	0.6	0.1	—	—	4.5	—	—	—	Rg	0.2	—	0.1	—	—	Rg
X̄		9.2	7.2						13.1	59.5		2.0				2.5					0.8			0.2		
s		8.8	3.4						7.4	14.5		2.0				1.9					0.6			0.4		
Kara-Tengir																										
KG-13	0.1—0.75	20.2	8.3	—	0.1	—	—	2.7	9.0	50.0	—	1.7	—	—	0.3	0.6	0.9	—	—	—	—	—	—	5.7	—	0.8
KG-35	0.1—0.75	25.7	12.8	—	—	—	—	1.5	15.4	34.5	Rg	0.5	—	—	—	3.8	0.8	0.2	—	—	—	—	—	4.7	—	—
KG-37	to 0.75	6.0	8.9	1.8	0.5	Rg	—	0.8	20.8	47.4	—	2.8	—	0.1	0.2	1.9	2.1	—	—	Rg	—	0.4	—	8.4	0.2	—
KG-38	to 0.5	22.0	11.7	1.8	Rg	—	—	Rg	15.8	39.8	—	0.1	—	—	0.2	6.7	0.4	—	—	—	—	—	—	1.1	—	—
KG-39	—	28.0	11.7	—	0.2	—	—	1.0	18.0	27.0	—	0.1	—	—	—	2.5	0.2	—	—	0.4	0.2	0.4	2.5	6.7	—	—
KG-40	0.1—0.7	30.0	5.1	—	Rg	—	—	0.7	16.6	37.8	—	1.0	—	—	—	1.8	—	—	—	—	0.3	0.6	—	5.3	—	—
KG-41	0.1—0.8	34.2	9.7	—	—	—	—	—	18.0	27.0	—	1.0	—	—	0.1	2.9	—	0.3	0.6	—	Rg	Rg	1.4	2.9	0.4	—
KG-42	0.02—0.5	13.4	5.6	—	0.8	—	0.2	1.9	17.5	30.9	—	3.1	—	—	—	7.0	8.0	—	—	—	0.2	3.2	—	10.9	0.1	—
KG-48	0.02—0.8	20.4	8.4	—	0.2	—	—	0.8	15.4	36.1	0.1	5.1	—	—	—	3.6	5.0	—	—	—	0.3	0.5	0.1	1.4	—	—
KG-59	—	16.4	7.3	1.8	0.9	0.1	—	2.4	23.8	19.2	0.9	9.8	—	—	—	2.5	—	—	—	—	0.5	—	—	24.3	—	—
KG-60	0.05—0.5	50.6	7.5	—	Rg	—	—	1.4	11.6	18.0	—	2.2	—	—	—	4.1	—	—	—	0.2	0.2	0.4	—	4.1	—	—
KG-69	to 0.7	29.3	6.3	—	0.1	—	—	0.7	13.8	30.7	—	1.2	—	—	—	4.8	0.4	—	—	0.2	0.3	—	Rg	10.6	1.3	—
KG-70	0.05—0.8	11.0	10.2	—	0.2	—	—	0.9	20.3	43.3	—	0.4	—	—	—	4.1	0.5	—	—	—	Rg	—	—	2.4	—	2.0
KG-71	—	6.8	8.8	4.2	—	—	—	Rg	11.7	55.8	—	0.1	—	—	—	11.3	—	—	—	0.2	Rg	—	—	0.4	—	4.7
X̄		22.4	8.7						16.3	35.5		2.1				4.1					0.1			6.3		
s		12.9	2.1						4.0	11.1		2.6				2.7					0.02			3.6		

TABLE 45. Values of Measured ρ_λ for Rocks from Red Beds of West Turkmenia
(Syrtlanli, Belek, and Kara-Tengir Sections)

No. of sample	Spectral luminance factors (%) for each wavelength											
	428	447	467	479	507	518	539	565	577	593	629	641

Syrtlanli

BAG-89	10	12.5	13.5	12.5	12.5	14	14.5	17.5	20.5	21.5	22	23.5
BAV-279	13.5	16.5	17	16.5	16	18.5	18.5	20.5	21	21	21.5	22.5
BAV-36a	11.5	16.5	16	19	19	19.5	19.5	23.5	24.5	25	24.5	25
BAV-36b	15	17	18	17.5	18	19.5	20	22.5	24	24	25	25
BAV-36c	11.5	14.5	16.5	17.5	17.5	18.5	19	24	24.5	24.5	25	25
BAV-42c	15.5	17	19.5	18.5	18.5	21.5	21	23.5	24	24	24.5	25.5
BAV-52	15.5	18	18	18	18	19.5	20	21.5	22.5	22	22.5	22.5
BAV-54	19	21.5	23.5	24	23.5	25.5	26.5	27.5	28.5	29	29.5	29.5
BAE-41	10	12.5	12.5	13	13.5	14	14.5	17.5	20.5	23	24	24
BAE-44	12	17	17.5	17	16	19.5	19.5	23	24.5	25	26	26

$\overline{X}$	13.4	16.3	17.2	11.4	17.3	19.0	19.3	22.1	23.5	23.9	24.5	24.9

S	2.9	2.7	3.1	3.2	3.1	3.3	3.4	3.0	2.4	2.3	2.3	2.0

Belek

B-3	15	14	16.6	16.3	14.6	15	16	27	29	29.6	33	32.3
B-4d	24.3	21.6	25	25	21.6	25.3	35	37	39.6	40.3	41.3	42
B-5	19.6	14.6	18.3	18	16.3	15	21.3	28	33.3	34.3	37	36.6
B-6b	16.3	16.3	19.3	17	16.3	18	19.3	28.3	29.3	32.3	36.6	37.6
B-7	17.3	16.6	15	15.6	14.3	15.3	16	22	22.3	25.3	26.3	29
B-9	19.6	16	17.6	17	15.6	15.6	17.6	25.6	29.6	30	32.6	33.3
B-14	18	15.6	17.6	18.3	17.6	17.3	18.6	25	29.6	32.6	34
B-15	16.3	14	15.3	16	16	15.3	17	22.6	25.6	28	30	31.3
B-17	14.3	13.6	15.3	16	15	15	16.6	22.3	24.6	26	27.6	28
B-19	14.6	15.3	17.6	17.3	16	16	16.6	24.3	26	27.6	31.6	32.6
B-21	15.6	15	16.6	17	17.6	17	19.6	24.3	26	29	29.6	31.3
B-23	13.6	12	14.6	15	14.6	14.3	15.3	20.3	21.6	25	26	27.3
B-24	11.6	14	15.6	15.3	15	15.3	17	24	26	27.3	29.6	30
B-26	17	14.3	16.6	18.3	18.3	17.3	20	25.6	30	33.3	35.5	37.3
B-27	25.3	22.3	26	26.6	26	26	27.6	34.6	38.3	40.3	43	44.5
B-28	20.6	17.3	18.3	20	17.6	18.3	20.3	28	28.3	32.6	34	34.3

$\overline{X}$	17.4	15.8	17.8	18.0	17.0	17.3	19.6	26.2	28.7	30.7	32.9	33.8

S	3.7	2.7	3.3	3.3	3.0	3.5	3.1	4.4	5.0	4.7	4.9	4.8

TABLE 45 (cont'd). Values of Measured ρ_λ for Rocks from Red Beds of West Turkmenia (Syrtlanli, Belek, and Kara-Tengir Sections)

No. of sample	Spectral luminance factors (%) for each wavelength											
	428	447	467	479	507	518	539	565	577	593	629	641

Kara-Tengir

No. of sample	428	447	467	479	507	518	539	565	577	593	629	641
KT-13	16	15.6	23	20.6	20.3	19.3	22.3	31.3	31.3	35	36.3	36
KT-35	10.3	11	14	13.6	15.3	14.3	16.6	24	25	26.3	26.6	26.6
KT-37	13	14.6	24.6	24.6	22	21.3	25.3	33.6	35	36.3	35.6	37
KT-38	14.3	13.3	18	17.6	17.3	15.6	19.3	25.6	28	30.3	33.3	33.3
KT-39	16	16.6	20.3	21.6	20	19	21.3	28	30	31.3	31.6	34
KT-40	16	15.3	17	17.6	16	14.3	17.3	22.6	26	27.3	23.6	30.6
KT-41	16.6	16.3	18.6	20	17.6	17.3	18	24.6	26.5	28	39.3	32
KT-42	17	16.3	17.3	19	18	17.3	18.3	25.3	26	29.3	33	33.6
KT-48	17.3	14.6	17	15.3	15	14.6	17.6	24	25.6	26.3	29.6	30
KT-59	15	17.6	20.6	20.3	19.3	20	22.3	29.6	32	34.6	35.6	35.3
KT-69	13.3	12.3	15.6	16	15.3	14.3	14.6	20.3	21.6	24	25	25.3
KT-70	18	16.3	19.3	19.6	19.3	19.3	21	27.3	30.3	33.3	34	34
KT-71	13.3	14	17.3	16.6	14.6	15.3	17	23	25.6	28.3	29.6	29.6
X	15.1	14.9	18.7	18.6	17.7	17.1	19.3	26.1	27.9	30.0	31.5	32.1
S	2.2	1.9	2.9	2.9	2.4	2.5	3.0	3.7	3.6	3.8	3.6	3.5

TABLE 46. Values of Polynomial Coefficients of ρ_λ Curves of Rocks from Red Beds of West Turkmenia (Syrtlanli Section)

No. of sample	a_0	a_1	a_2	a_3
BAG-2/122	18.92	1.086	−0.03477	−0.05315
BAG-5	21.00	1.097	−0.02902	−0.04301
BAG-24	20.18	1.216	−0.06770	−0.00892
BAG-44	19.19	1.085	0.01074	−0.05752
BAG-89	16.86	1.197	0.05165	−0.08339
BAG-94/98	18.88	1.238	0.00425	−0.11206
BAG-141/110	16.85	0.957	−0.01828	−0.04650
BAG-155	12.18	0.911	0.01958	−0.08741
BAG-191	17.58	1.216	0.01978	−0.14528
BAG-200	18.62	1.031	−0.07557	−0.01031
BAV-74/92	17.68	1.030	−0.06149	−0.07185
BAD-1	18.75	1.324	0.00220	−0.10490
BAD-16	20.64	1.020	−0.02577	−0.07028
BAD-84	18.36	1.288	−0.00195	−0.06591
BAD-98	22.54	0.947	−0.02812	0.02098
BAD-109/155	21.400	0.971	−0.04056	−0.08182
BAD-111	22.29	1.345	−0.01998	−0.07343
BAD-117	14.05	1.431	0.03546	−0.11031
BAD-132	22.88	0.929	−0.04725	−0.02343
BAE-1	19.98	1.185	0.00509	−0.10629
BAE-41	17.09	1.298	0.03556	−0.10594
BAE-42	21.88	0.976	−0.04341	−0.02780
BAE-44	20.69	1.124	−0.02013	−0.05892
BAE-75	18.99	0.874	−0.00225	−0.02657
BAE-78	18.36	0.798	0.01044	−0.02203
$\bar{X}$	19.03	1.103	−0.01286	−0.06304
S	2.525	0.1654	0.03329	0.03972

TABLE 47. Values of Polynomial Coefficients of P_λ Curves of Rocks from Red Beds of West Turkmenia (Belek and Kara-Tengir Sections)

No. of sample	a_0	a_1	a_2	a_3
Belek				
B-3	19.658	−3.9218	0.7011	−0.02418
B-4d	28.852	−6.083	1.185	−0.05008
B-5	27.338	−6.940	1.131	−0.0411
B-6b	18.252	−2.481	0.437	−0.0096
B-7	19.428	−2.822	0.373	−0.008
B-9	24.233	−4.866	0.793	−0.0284
B-14	21.798	−3.396	0.560	−0.017
B-15	19.928	−3.246	0.518	−0.015
B-17	18.094	−3.117	0.575	−0.021
B-19	12.342	+0.610	−0.035	+0.010
B-21	17.829	−2.025	0.408	−0.013
B-23	11.274	−0.967	0.284	−0.007
B-24	13.072	−0.967	0.299	−0.009
B-26	20.435	−3.236	0.564	−0.016
B-27	28.653	−3.172	0.565	−0.017
B-28	26.067	−4.870	0.769	−0.026
$\overline{X}$	20.453	−3.219	0.5704	−0.01827
S	5,517	1,914	0.3062	0.01408
Kara-Tengir				
KT-13	17.486	−1.155	0.376	−0.013
KT-35	12.343	−1.924	0.536	−0.023
KT-37	10.880	+1.654	0.1295	−0.007
KT-38	17.0299	−1.7680	0.3775	−0.0108
KT-39	21.6572	−5.5768	1.1739	−0.0519
KT-40	28.936	−9.5462	1.6201	−0.0688
KT-41	16.953	−0.2514	0.0623	+0.00408
KT-42	23.389	−4.833	0.820	−0.03001
KT-48	21.9988	−4.0762	0.6346	−0.0206
KT-59	16.525	−1.0483	0.380	−0.0137
KT-69	14.175	−0.787	0.184	−0.0039
KT-70	21.948	−3.520	0.671	−0.0247
KT-71	15.654	−1.672	0.361	−0.0109
$\overline{X}$	18.383	−2.6541	0.5635	−0.02109
S	4.992	2.865	0.4378	0.01996

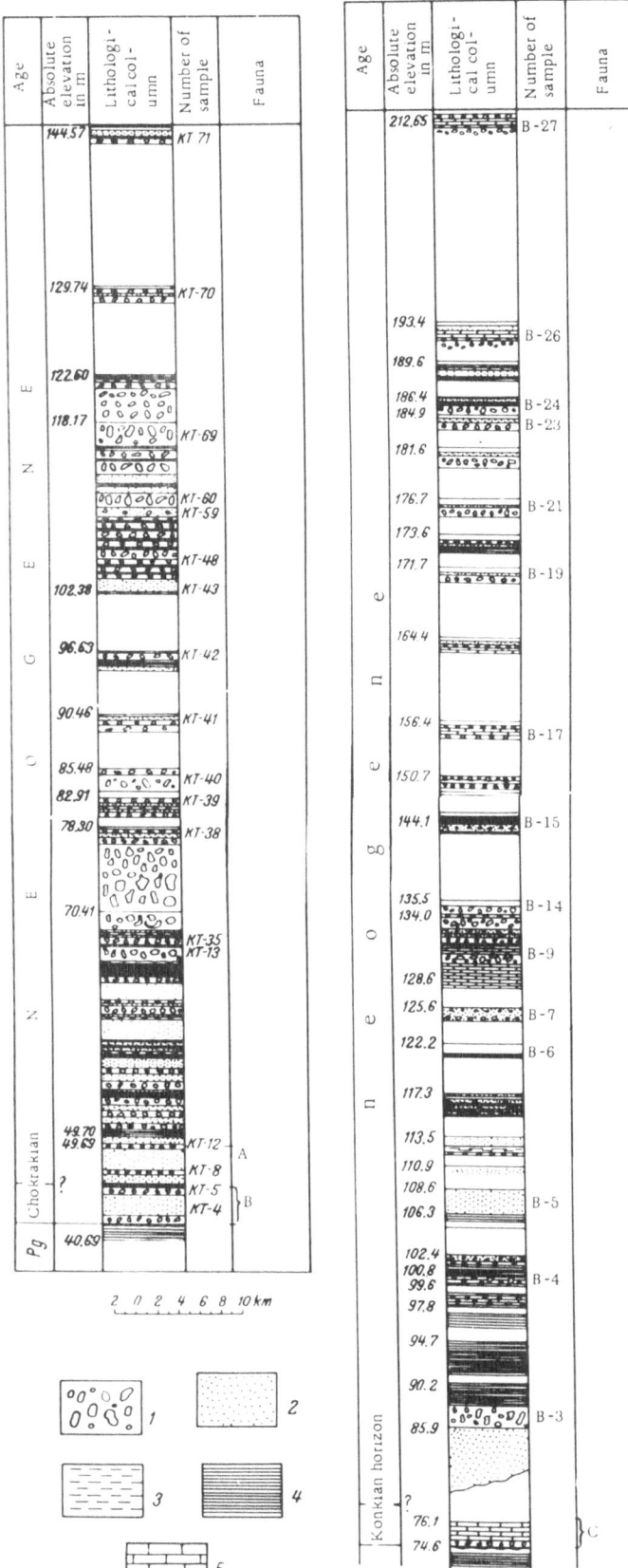

The high barite content (33%) is characteristic.

Sample B-14 – from a bed of conglomerates containing large pebbles, mainly of granitoid composition. Magnetite and limonite predominate in the heavy fraction, which also contains a large amount of zircon. The limonite consists of pseudomorphs of wood, plant roots, and pyrite. The magnetite consists of small shining crystals, often of octahedral shape.

Sample B-15 – from a bed of fine-pebble conglomerates, alternating with bands of coarse-grained sands. The largest and subangular pebbles consist of granites and granite-porphyries. The heavy fraction contains a large amount of coarsely crystalline magnetite and ilmenite. Zircon and celestite are characteristic.

Sample B-17 – from a bed of conglomerate containing sparse large limestone pebbles and abundant smaller pebbles of granites and felsite porphyries; the cement is sandy. The heavy fraction consists mainly of magnetite and contains a large amount of short prismatic zircon crystals, which often contain air bubbles.

Sample B-19 – from a bed of conglomerate containing semiangular pebbles of granites, very angular pebbles of porphyrites and porphyries, and well-rounded limestone pebbles. Magnetite and thickly tabular ilmenite predominate in the heavy fraction

---

Fig. 79. Stratigraphic sections of red beds in Kara-Tengir and Belek regions (Krasnovodsk Peninsula). 1) Conglomerates; 2) sands; 3) silts; 4) clays; 5) limestones. A*) Clauselidae, Helicidae; B) Pecten pertinax Zhizh., Leda pella L., Ervilia praepodolica Andrus., Cardium ex. gr. multicostatum Brocc, C. impar Zhizh., Ostrea sp., Modiolus marginatus Eichw., Donax sp.; C) Corbulla gibba Olivi, Ostrea digitalina Dub., Anemia ephippium L., Chlamys malvinae Dub., Pitar islandicoides Lamarck, Anadara diluvii Lamarck, Chione multilamella Lamarck, Ch. basteroti Desh, Nuculana sp., Cardium sp., Natica sp., Timoclea (Parvikenus) konkensis Sov., Ervilia trigonula Sokol., Mactra (Eomactra) basteroti Mayer., Anadara turontensis Duj. var. konkensis.

*Identification of fauna by Korobkov.

121

obtained by washing off the sandy cement from these conglomerates, and short prismatic crystals of zircon are numerous.

Sample B-21 – from a bed of coarse-pebble conglomerate in which the pebbles reach the size of boulders (up to 54 cm) and consist mainly of granites and granitic dike rocks. Limestones comprise about 5%. The heavy fraction contains a large amount of magnetite and zircon; celestite is characteristic.

Sample B-23 – from a coarse-pebble conglomerate with coarse-grained sandy cement. The pebbles consist mainly of granites and granite dike rocks. Magnetite again predominates in the heavy fraction.

Sample B-24 – from a bed of fine-pebble conglomerate with loose sandy cement. The percentage of granitic and porphyrite dike rocks is increased at the expense of the granites. As in the previous sample the heavy fraction contains a large amount of magnetite and a fair amount of zircon. A characteristic feature is the presence of large, pale brown crystals of sphene.

Sample B-26 – from a bed of conglomerates with clayey-sandy cement. The pebbles are composed mainly of large-grained red granites and granitic and porphyrite dike rocks. The heavy fraction contains up to 75% magnetite in the form of large shining crystals, often in aggregates with quartz, epidote, and sphene. The sphene is very characteristic; its crystals are envelope-shaped, up to 1 mm in length, and greenish brown or pale brown. Aggregates of limonite with rock are numerous.

Sample B-27 – from a bed of conglomerate with clayey cement. Here again the amount of limestone pebbles is increased at the expense of the large-grained granites. The heavy fraction contains about 70% magnetite in the form of octahedral crystals and aggregates with transparent quartz; the magnetite grains are shiny and large – up to 0.2-0.3 mm. There is a large amount of titanium minerals – anatase, rutile. A characteristic mineral is celestite, which occurs in the form of large prismatic bluish transparent crystals ($Ng = 1.630$) up to 2 mm long.

Sample B-28 – from a hole dug on the plateau. This sample completes the Belek section. About half of the pebbles of the conglomerate are granites, one third are porphyrites and porphyries, and about 17% are limestones. The heavy fraction from the sandy cement of these conglomerates is very rich in magnetite, the large crystals of which reach 1 mm across. Signs of solution can sometimes be observed on the faces of the magnetite crystals. Pseudomorphs of limonite from pyrite occur. Cryptocrystalline aggregates of celestite are characteristic.

<u>Section of Red Beds in Kara-Tengir Region.</u> The section begins at the base of the souther scarp of Kyuryanyn-Kyure, 52 km east of Krasnovodsk and approximately 2 km north of Kara-Tengir. The line of the section, which consists of a series of prospect holes and strippings, runs in the direction N 55° W.

At the base of the section the deposits of calcareous sandstones with bioherms and inclusions of pebbles mostly of granitic composition are overlain by coarse-grained sandstones separated from the Chokrakian deposits by an interlayer of marly, highly gypseous limestone, 0.25 m thick.

Higher up, the section conglomerates with large polymictic pebbles and poorly sorted sandy cement predominate. A general view of the section is shown in Fig. 79 (KT). The places from which the samples were taken from this section are marked with numbers.

The petrographic-mineralogical composition of samples taken from the Kara-Tengir section can be briefly characterized as follows (from the bottom up):

Sample KT-8 – from a bed of fine-pebble conglomerate with sandy-clayey cement. The pebbles contained in this conglomerate do not exceed 5 cm across. Granites and granophyric dike rocks are noticeably predominant among the pebbles; limestone and marl pebbles make up about 15% of the volume. The heavy fraction washed from the conglomerate cement consists of about 33% magnetite. Isolated grains of bismuthinite are present. The top of these conglomerates contains rusty brown sandstones containing fine pebbles. The total thickness of the bed is 2.10 m.

Higher up the section there are coarse-layered fine-pebble conglomerates with an appreciably higher proportion of rounded limestone pebbles; the granite and felsite-porphyry pebbles in this bed are distinguished by their angularity. The heavy fraction from the cement of these conglomerates consists mainly of grains of ore minerals: magnetite and ilmenite colites. Relatively large zircon crystals are present.

Sample KT-13 – from a bed of large-pebble conglomerate. The boulders reach a size of 0.56 m. The largest of them consist of pink crystalline limestones and there are many porphyrites. The amount of granite pebbles is noticeably smaller, but the granite-porphyry dike rocks chiefly remain. Half of the heavy fraction consists of magnetite in the form of large angular crystals (octahedra) with mat faces. Sometimes magnetite is found in aggregates with quartz. Celestite is present and, as in the previous sample, there is a large amount of limonite.

Sample KT-35 – from fine-pebble conglomerates with sandy cement alternating with bands of fine-grained sandstones. The pebbles consist mainly of large limestone boulders, which make up 50% of the total number of pebbles. Dike porphyries and porphyrites are numerous. The heavy fraction from the sandy cement of these conglomerates consists mainly of ore minerals: magnetite, ilmenite, hematite, and limonite, which constitute almost 90% of the whole heavy fraction.

Sample KT-37 – from a bed of large-pebble conglomerates with sandy cement. The largest pebbles reach 0.65 m. The boulders and pebbles consist mainly of crystalline limestones, porphyrites and porphyries. Rare boulders of biotite granites up to 0.40 m in diameter occur. The heavy fraction consists mainly of magnetite in the form of angular grains and ilmenite. Celestite is present.

Sample KT-38 – from a bed of coarse-grained sand separating thick large-pebble conglomerates. The heavy fraction from this sample consists mainly of ore minerals: magnetite, ilmenite, limonite, and hematite.

Sample KT-39 – from a bed of large-pebble conglomerate with sandy cement. The large pebbles (10-12 cm) are represented by crystalline limestones, sometimes oolitic, and there are many porphyrites. The heavy fraction from the sandy cement of these conglomerates contains many ore minerals: magnetite, ilmenite, hematite, and limonite, which together make up about 80% of the heavy fraction. A characteristic feature is the presence of native bismuth in the form of soft lamellar pale gray grains, which effervesce vigorously in HNO_3 and give a positive reaction with potassium iodide. There is a large amount of barite and celestite.

Sample KT-40 – from a thick bed of fine-pebble conglomerates containing large boulders of limestones among which pink crystalline limestones predominate. There are many calcareous sandstones and porphyrites. The heavy fraction consists mainly of ore minerals: magnetite, limonite, and ilmenite.

Sample KT-41 – from a bed of large-pebble conglomerate with sandy cement. The pebbles consist mainly of pink limestones, crystalline or chalklike. There are many porphyrites and the number of granite pebbles is again increased. The heavy fraction consists mainly of ore minerals: limonite, magnetite, ilmenite, hematite, and a considerable amount of celestite.

Sample KT-42 – from a bed of large-pebble conglomerate, the pebbles of which consist of rounded boulders of limestones and porphyrites. The heavy fraction is characterized by a predominance of large (up to 0.5 mm) round grains of limonite with conchoidal cleavage. The amount of magnetite is reduced, but is still fairly high (31%).

Sample KT-48 – from a bed of fine-grained clayey sandstone. The heavy fraction from this bed, the material of which is quite unsorted and almost unrounded, consists mainly of iron minerals: magnetite, limonite, ilmenite, and hematite.

Sample KT-59 – from coarse-grained sand containing large pebbles of crystalline and chalklike limestones, plagioclase porphyrites, and porphyries. The heavy fraction contains a great amount of magnetite, ilmenite, and limonite and is characterized by an increased content of zircon and celestite.

Sample KT-60 – from a bed of large-pebble conglomerate with sandy cement. Among the pebbles there are numerous white chalklike limestones, which reach 20 cm across. The heavy fraction contains limonite (more than 50%), magnetite, and a large amount of celestite and zircon.

Sample KT-69 – from a bed of fine-pebble conglomerate. The largest pebbles consist of porphyrites, up to 8 cm across, and they make up 35-40% of the total number of pebbles. There are many granite pebbles, mainly coarse-grained, and sometimes greisenized. The heavy fraction from the sandy cement of these conglomerates consists mainly of magnetite in the form of angular grains up to 0.7 mm across and has a high content of coarsely crystalline celestite.

Sample KT-70 – from coarse-grained sand cementing a conglomerate consisting of pebbles of granitoids (up to 35%), porphyrites and porphyries (up to 36%), and sedimentary rocks in the form of sandstones and silty pelites. The pebbles reach 12 cm or more across. Approximately half of the heavy fraction from this sand is made up of magnetite, ilmenite, hematite, and limonite, which together account for more than 45%, and there is also a large amount of celestite.

Sample KT-71 – from a conglomerate consisting mainly of pebbles of granite and porphyrites. The largest pebbles, up to 15 cm across, consist of coarse-grained biotite and hornblende granites. The proportion

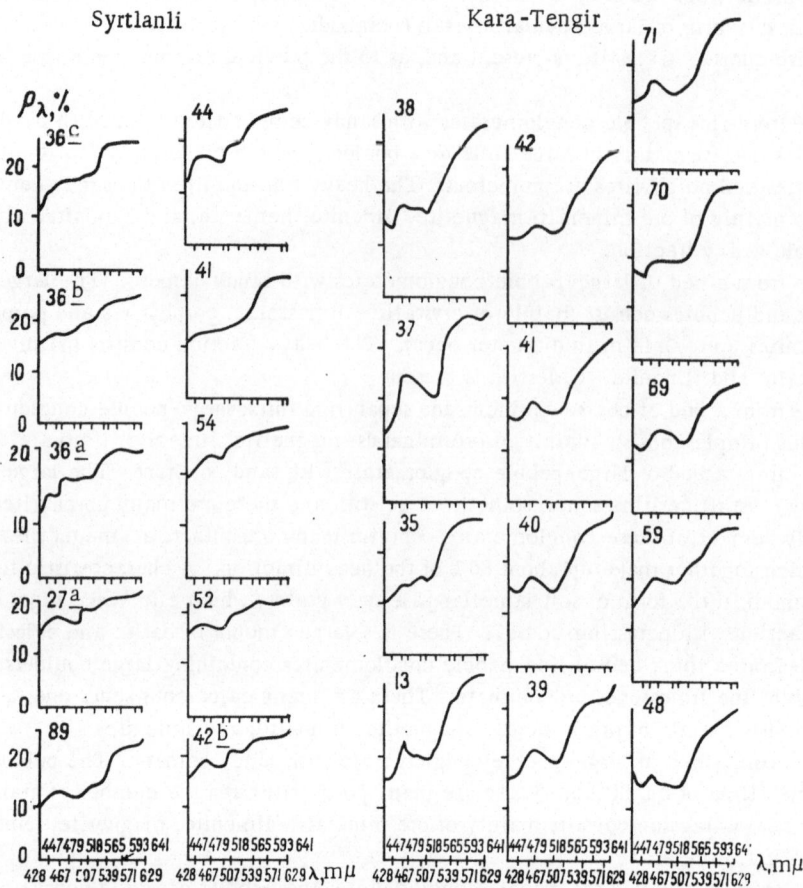

Fig. 80. Nature of ρ_λ curves of samples from compared sections (Syrtlanli and Kara-Tengir) of red beds of West Turkmenia.

of limestone and marl pebbles does not exceed 4%. In the minerals of the heavy fraction the amount of magnetite is 56%. The grains are large, of lamellar habit, up to 0.6 mm, and often occur in aggregates with quartz, feldspar, and sphene.

Above lies the surface of the plateau. The thickness of the soil layer reaches 30 cm.

As the description shows, samples from the compared sections of Neogene deposits formed in conditions close to typical geosynclinal (Syrtlanli section) and platform conditions (Belek and Kara-Tengir sections) differ appreciably in lithological composition and external appearance.

While coarse-grained sands and brick-red clays predominate in the Syrtlanli section, conglomerates with sandy cement alternating with brown clays predominate in the Belek and Kara-Tengir sections.

The mineralogical composition of samples taken from the sandy formations of these deposits differs appreciably as regards the mineral complex of the heavy fraction. The heavy residues from samples of the Syrtlanli section are rich in granite, epidote, and ilmenite (see Table 44). The heavy residues from samples of the Krasnovodsk sections have approximately the same granulometric composition (sands), are very rich in magnetite, limonite, and characteristic formations in the form of barium and strontium sulfates (see Table 44).

The results of photometry of the samples from the compared sections of Neogene strata are given in the form of ρ_λ values in Table 45, and the same data are illustrated in the form of ρ_λ curves in Fig. 80. The difference in the character of the ρ_λ curves from these sections is so obvious that one can say at a glance whether the curve belongs to a rock from the Syrtlanli section or from the Kara-Tengir section.

124

TABLE 48. Contingencies for Polynomial Coefficients of ρ_λ Curves for Sections of Red Beds of West Turkmenia

Region	$a_{10} \leqslant 0$	$a_{10} > 0$	Σ
Kara-Tengir and Belek	$\dfrac{27}{14.500}$	$\dfrac{2}{14.500}$	29
Syrtlanli	$\dfrac{0}{12.500}$	$\dfrac{25}{12.500}$	25
Σ	27	27	54

$$\chi^2 = 46.6; \ P(\chi^2) < 0.001.$$

Region	$a_{20} \leqslant 0.045$	$a_{20} > 0.045$	Σ
Kara-Tengir and Belek	$\dfrac{1}{13.426}$	$\dfrac{28}{15.574}$	29
Syrtlanli	$\dfrac{24}{13.3400}$	$\dfrac{1}{13.426}$	25
Σ	25	29	54

$$\chi^2 = 46.3; \ P(\chi^2) < 0.001.$$

Region	$a_{30} \leqslant -0.025$	$a_{30} > -0.025$	Σ
Kara-Tengir and Belek	$\dfrac{8}{15.574}$	$\dfrac{21}{13.426}$	29
Syrtlanli	$\dfrac{21}{13.426}$	$\dfrac{4}{11.574}$	25
Σ	29	25	54

$$\chi^2 = 17.2; \ P(\chi^2) < 0.001.$$

The results of smoothing of the ρ_λ curves of sand samples from these two regions are given in Table 46 and 47 and are shown in the form of frequency distributions of the values of the polynomial coefficients in Fig. 81. The nature of the obtained histograms shows that the samples can be distinguished from the values of the polynomial coefficients a_1, a_2, and a_3. A point worth stressing is the uniformity of the distribution of the values of a_1 and a_2, which have a narrow range of variation.

The results of a χ^2 test of the difference in the values of the polynomial coefficients for the compared sections are given in Table 48. As the obtained values of χ^2 show, the differences in the photometric parameters of sand samples from the Syrtlanli section and samples of the sand fractions from the Belek and Kara-Tengir sections are highly significant.

Thus, a test of the method in this third case confirms that deposits formed in different facies conditions of sedimentation can be distinguished by their photometric parameters or, more precisely, from the curves of the spectral luminance factors of photometered samples from different sections.

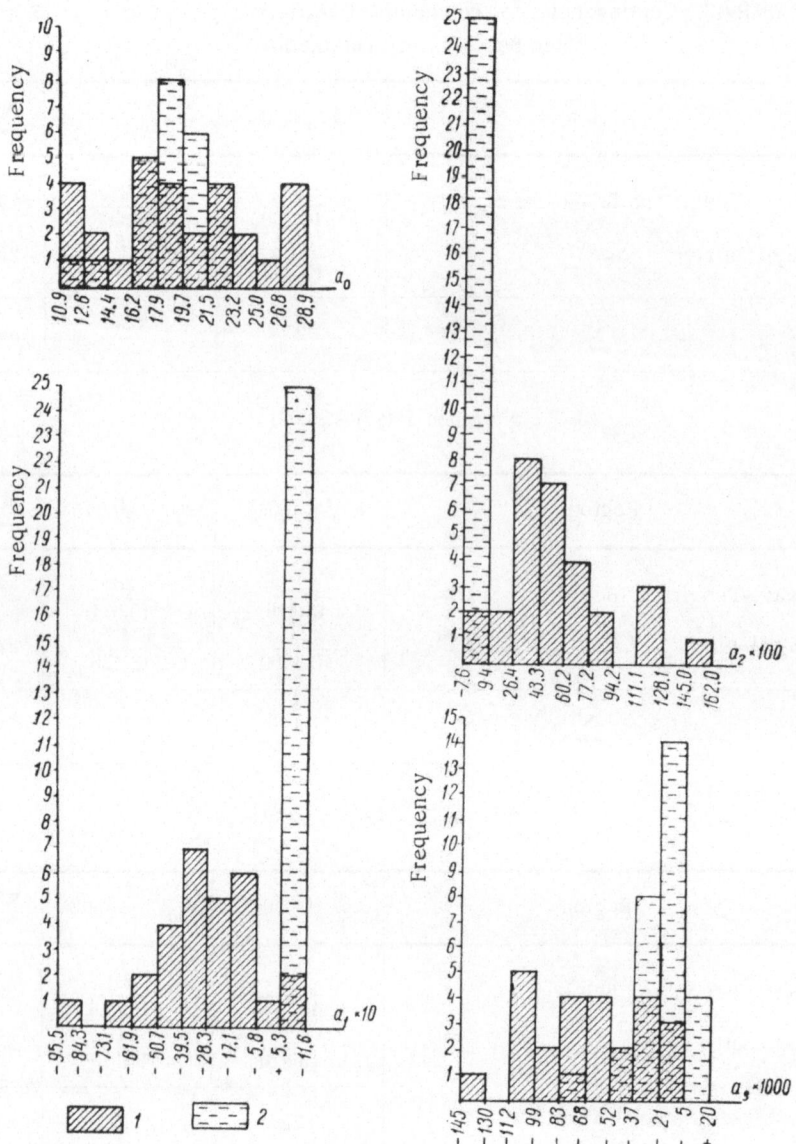

Fig. 81. Frequency distribution of values of polynomial coefficients of ρ_λ curves of rocks from red beds of West Turkmenia. 1) From Kara-Tengir section; 2) from Syrtlanli section.

Summing up the data of the laboratory stage of measurement of spectral luminance factors of rocks, as exemplified by the terrigenous deposits of West Turkmenia, we can draw the following conclusions:

1. Deposits formed in different facies environments have their own characteristic photometric parameters, although the rocks may resemble one another externally and in petrographic composition. This is confirmed by samples of sandy-silt composition from Middle Jurassic, terrigenous Cretaceous, and Neogene deposits.

2. The photometric properties of rock samples from different regions, but from synchronous deposits of similar composition, is affected by the amount of soluble ferrous or ferric iron in the rock, by the carbonate or clay content, and sometimes by the content of the characteristic minerals of the heavy fraction.

3. By measurement of the spectral luminance factors of rocks and evaluation of the ρ_λ curves it is possible to class the investigated deposits into lithofacies types, i.e., to carry out lithological mapping.

CHAPTER VII

AN AEROPETROGRAPHIC SURVEY OF MODERN SANDY DEPOSITS

The association discovered in the laboratory between the spectral luminance of sandy-silt deposits and their lithofacies characteristics had to be verified in the field – from the air. For this purpose we conducted the first experimental aerial spectrophotometric survey of sands in August 1957. The work was conducted on an area adjoining the northwest shore of the Caspian Sea, in the region of the Kuma, Terek, and Sulak deltas.

Below we give the results of this survey as a demonstration of the possibility of aeropetrographic mapping of sandy deposits. Some of the data given have been published before in the form of separate papers (Romanova, 1958, 1960a, 1960b).

This experimental aeropetrographic survey was conducted on the territory of northeast Ciscaucasia roughly within the triangle formed by the towns of Makhachkala, Budennovsk, and Astrakhan. This region contains sands of different petrographic composition: almost monomictic quartz sands, occupying the northern part of the experimental polygon in the region of the Volga delta, and typical graywacke sands, occupying the southern part of this area in the region of the Terek and Sulak deltas.

The spectrophotometric survey within the experimental polygon was performed with the RShch-1 aerial cinespectrograph. The aircraft used for the survey were the YaK-12 and YaK-12a light airplanes. A height of 20 m was adopted as the working height during the survey. Owing to the uneven surface of the surveyed barchans, the height of the survey varied from 10 to 20 m. The direction of the flights over the object depended on the direction of the wind and the position of the airplane's shadow. Photometry was restricted to the period between 10 AM and 2 PM local time. The survey procedure began and ended with photometry of the standard attenuated in varying degree. In one case we used attenuating filters consisting of a set of photographic plates with different photographic density equal to 0.30, 0.52, 0.61, 0.90, 1.00, and 1.55 density units. The uniformity of the densities and their values were checked by repeated measurements with a densitometer. In the photometry of the standard, the filters were mounted in a special holder directly in front of the objective.

The spectrophotometry was carried out with cinefilm of AM-35 type. The film speed was 24 frames per sec and the width of the spectrograph slit was 0.193 or 0.142 mm.

The exposed film was developed in D-76 metol-hydroquinone developer or Chibisov developer. In view of the climatic conditions in the deserts the film was fixed in acid hardening fixer. The processing of the exposed film was controlled by superimposing a sensitogram at the beginning and end of each film.

The spectrograms were measured on the MF-4 recording microphotometer or the MF-2 microphotometer. From the series of spectrograms obtained in the survey of an outcrop we selected for measurement three frames in which the surface of the photometered area was cleanest and of approximately the same size. Spectrograms of the tops of flat barchans were chosen. From the mean values of ρ_λ obtained by measurement of the spectrograms of the investigated objects we plotted the curves of ρ_λ. The curves were numerically evaluated by expansion in an orthogonal series with the aid of Chebyshev polynomials to the cubic term of the polynomial.

For a petrographic-mineralogical characterization of the photometered sands, we collected sand samples from each object where it was possible to land the airplane at a site selected from the air. From the collected samples of 2 kg we quartered about 10 g for granulometric analysis and approximately 20 g for petrographic analysis. From the rest of the sand we obtained the heavy residue by triple washing.

The composition of the investigated rocks was determined by granulometric, petrographic, and mineralogical analyses with a count of grains in the heavy residue and thin sections.

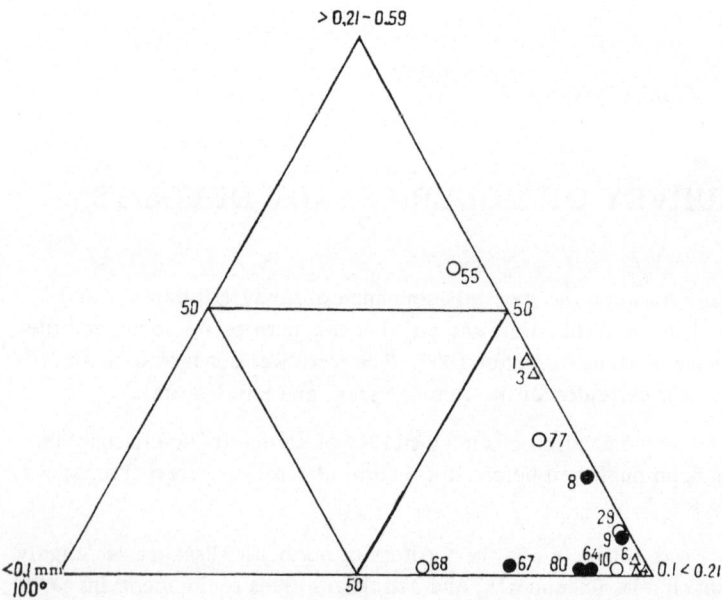

Fig. 82. Granulometric composition of sand samples from Northwest Caspian region. Δ) Sulak; ○) Terek; ●) Volga types.

The granulometric analyses of the sands were performed by a simplified Williams method, but we dispensed with elutriation in view of the absence or very insignificant amounts of clay. Thus, the fraction less than 0.001 mm was included in the fractions of 0.05 mm and less.

The spectrophotometric data were compared with the results of the petrographic-mineralogical analyses by the χ method and correlation analysis.

Within the northwest Caspian region we photometered about a hundred outcrops, the locations of which are shown by circles in Figs. 84 and 85. The figures in the circles correspond to the number of the photometered object and the number of the sample taken from this object for petrographic-mineralogical analysis. The survey flights were approximately longitudinal, so that one film carried the results of photometry of sands of different petrographic types.

The distribution of the petrographic types of sands within the experimental polygon is indicated by isolines. One of the isolines, corresponding to an ilmenite content of 40% in the heavy residue, marks the southern limit of the occurrence of predominantly quartz sands resembling the typical monomictic sands from the Volga delta, and, hence, these sands were called the Volga type. The isoline for a 10% content of orthorhombic pyroxenes in the heavy residue marks the lower limit, southwest of which lies the area occupied by Terek sands.

In the southeast of the polygon – in the Sulak delta – the sands are of graywacke composition and contain numerous fragments of effusive rocks and shales. These are the Sulak type of sands.

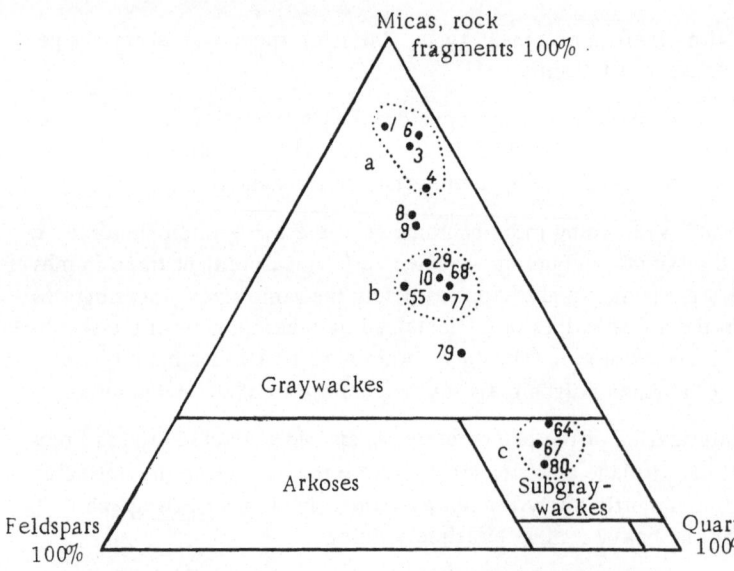

Fig. 83. Petrographic composition of sands (classification of Dapples, Krumbein, and Sloss). a) Sulak; b) Terek; c) Volga.

Sands of these three types have different granulometric and petrographic composition, different degrees of rounding and sorting of the fragmentary material, and, also, as will be seen later, different reflectance.

The investigated sands of each of the distinguished types are described below.

Sands of Sulak Type

The results of the granulometric analyses showed that the coarsest-grained sands occur in the Sulak delta, where the 0.15- to 0.21-mm fraction predominates. They are very similar to sands of the Terek type in grain size and degree of sorting. Sands of the Volga type are finer grained, the predominant fraction

128

TABLE 49. Petrographic Composition of Sands of Northwest Caspian Region

No. of sample	Sand type	Rock-forming minerals			Rock fragments		
		quartz	feldspars	dark-colored minerals	shales, siltstones	volcanic fragment	lime-stones
1	Sulak	8.5	11.6	1.2	20.1	18.0	40.6
3		13.8	9.6	4.5	10.7	22.9	38.5
4		21.2	10.1	9.3	15.7	18.8	24.9
6		15.3	6.8	4.9	8.6	26.8	37.6
10	Terek	32.3	16.1	10.3	3.4	27.9	10.0
29		29.0	15.8	8.5	8.0	27.9	10.8
55		27.4	22.1	7.3	11.7	18.5	13.0
68		33.6	13.0	8.5	5.6	30.1	9.2
77		34.6	16.8	8.5	13.0	18.6	8.5
64	Volga	64.8	9.7	6.9	8.8	7.0	2.8
67		66.4	13.9	6.5	1.4	6.6	5.2
80		69.4	15.4	2.4	3.6	6.9	2.3
8	Interme-diate com-position	20.2	13.6	10.7	12.8	25.3	17.4
9		21.6	13.4	16.7	6.1	34.7	7.5
79		42.7	19.1	7.3	7.8	16.0	7.1

being 0.10-0.15 mm, and they are better sorted (Fig. 82).

The degree of rounding of the fragmentary material is different in these types of sands. The grains of sands of the Sulak type are half-rounded or rounded and are often lenticular or flat and elongated. The grains of sands occurring along the Terek are half-rounded or angular. The grains of sands of the Volga type are usually almost round. The grains of sands on the shore of the Caspian Sea (outcrops 8, 9, 10, etc.) are appreciably more rounded.

The petrographic composition of the investigated sands is given in Table 49. Sands of the Sulak type are richest in rock fragments. They occupy the top corner of the graywacke field on the classification triangle of Dapples, Krumbein, and Sloss (Fig. 83). The fragmentary material consists mainly of brown limestones with a compact pelitic structure, and some contain the remains of foraminifera. Fragments of lithic and vitric tuffs and porphyrites are abundant. Fragments of pyroxene porphyrites are rarer, but there are numerous fragments of clayey shales, with ferric hydroxide present on the bedding planes. The amounts of quartz and feldspars are very small. The quartz grains are large, with mosaic extinction, and gas-bubble inclusions can be seen in some of them. Limonite is relatively abundant (for Sulak sands in Table 49 the column "dark-colored minerals" gives the amount of limonite fragments in the sands).

Ilmenite and limonite predominate in the heavy fraction of Sulak sands, and there is a large amount of zircon (Table 50). The distribution of these minerals is uneven; for instance, in certain cases the limonite content of the heavy fraction reaches 78.0%, whereas in other places only rare grains are present. A similar situation is observed in the case of zircon, the amount of which sometimes reaches 32.3% and sometimes only a fraction of one per cent.

The morphological characteristics of some minerals of the heavy fraction of Sulak sands are given below. The grain size in the heavy fraction is 0.05-0.75 mm. The bright luster of the grains and the different degrees of rounding are characteristic.

Corundum – slightly rounded or angular grains, colorless or mottled blue.

Ilmenite – usually rounded, rarely angular, grains of irregular shape, black in color, and with a shiny surface.

TABLE 50. Mineralogical Composition (%) of Heavy Residue of Sands of Northwest Caspian Region

No. of Sample	Corundum	Hematite	Limonite	Ilmenite	Rutile	Anatase	Leucoxene	Spinel	Chromite	Magnetite	Sillimanite	Kyanite	Staurolite	Zircon	Antho-phyllite	Amphiboles	Orthorhombic pyroxenes	Monoclinic pyroxenes	Garnets	Epidote	Olivine	Tourmaline	Chlorite	Sphene	Monazite	Apatite	Carbonates	Barite
1	2	3	4	5	6	7	8	9	10	11	12	13	14	15	16	17	18	19	20	21	22	23	24	25	26	27	28	29

Sulak type

No.	Corundum	Hematite	Limonite	Ilmenite	Rutile	Anatase	Leucoxene	Spinel	Chromite	Magnetite	Sillimanite	Kyanite	Staurolite	Zircon	Antho-phyllite	Amphiboles	Ortho-pyroxenes	Mono-pyroxenes	Garnets	Epidote	Olivine	Tourmaline	Chlorite	Sphene	Monazite	Apatite	Carbonates	Barite
1	—	2.8	78.0	2.5	1.8	—	1.1	Rg	2.5	Rg	—	Rg	0.7	0.7	—	1.1	1.1	0.7	1.1	5.2	—	Rg	—	—	0.4	Rg	—	Rg
2	Rg	7.4	3.8	22.9	3.5	0.1	1.5	Rg	1.0	17.0	Rg	Rg	0.3	4.9	Rg	0.7	20.7	0.3	12.4	2.3	—	Rg	—	Rg	Rg	Rg	—	1.2
3	—	3.4	24.9	31.7	4.6	Rg	4.2	0.2	1.4	Rg	—	0.1	0.9	10.5	Rg	0.5	4.3	0.2	6.4	3.1	—	0.2	—	—	Rg	0.1	0.2	3.1
4	Rg	11.6	0.9	32.1	3.8	Rg	1.8	Rg	3.0	Rg	—	Rg	Rg	32.3	—	—	0.2	Rg	6.0	0.2	—	Rg	—	0.6	Rg	Rg	—	7.9
5	Rg	9.8	2.7	43.4	3.8	0.3	4.6	—	2.2	Rg	—	0.5	0.5	17.2	—	0.3	0.8	Rg	7.9	2.7	—	Rg	Rg	0.2	Rg	—	0.8	1.9
6	Rg	Rg	Rg	36.6	6.7	1.3	3.6	Rg	1.0	16.5	—	—	0.6	5.5	—	0.6	5.7	1.4	13.9	2.4	—	Rg	—	0.2	Rg	0.7	—	3.3

Terek type

No.	Corundum	Hematite	Limonite	Ilmenite	Rutile	Anatase	Leucoxene	Spinel	Chromite	Magnetite	Sillimanite	Kyanite	Staurolite	Zircon	Antho-phyllite	Amphiboles	Ortho-pyroxenes	Mono-pyroxenes	Garnets	Epidote	Olivine	Tourmaline	Chlorite	Sphene	Monazite	Apatite	Carbonates	Barite
10	Rg	3.5	1.0	32.5	2.1	0.2	1.3	0.2	1.3	42.9	—	Rg	Rg	Rg	—	Rg	4.8	—	9.7	0.5	—	Rg	Rg	Rg	0.2	Rg	—	—
29	Rg	1.9	2.3	31.0	1.9	Rg	Rg	Rg	1.0	15.9	Rg	0.4	0.2	4.9	—	0.3	19.0	1.6	14.7	4.4	Rg	—	—	0.6	0.2	Rg	—	—
30	Rg	0.6	19.5	7.2	1.7	—	0.6	Rg	Rg	10.3	0.3	0.1	Rg	2.6	0.3	1.5	36.0	2.3	15.0	10.1	Rg	0.5	Rg	Rg	0.2	0.2	—	—
31	Rg	Rg	6.8	8.4	0.5	0.1	0.1	—	0.8	21.7	0.5	Rg	0.8	2.5	0.7	0.2	27.6	3.3	16.2	8.9	Rg	Rg	Rg	0.1	Rg	Rg	—	—
32	Rg	Rg	2.0	30.8	0.7	0.1	2.3	Rg	0.7	22.5	Rg	0.2	Rg	2.8	Rg	4.3	14.8	Rg	22.1	1.3	Rg	Rg	Rg	—	Rg	Rg	—	—
33	Rg	2.8	5.9	18.9	0.3	0.1	1.4	Rg	0.5	11.8	Rg	0.1	—	2.2	Rg	1.9	18.6	Rg	25.4	6.9	—	0.3	Rg	0.2	—	—	—	—
34	Rg	—	4.7	17.6	0.8	0.1	0.6	Rg	Rg	13.3	Rg	—	Rg	2.3	Rg	2.0	29.2	Rg	24.5	6.6	—	Rg	Rg	Rg	Rg	Rg	—	—
35	Rg	9.5	1.1	27.2	1.0	Rg	1.3	—	0.3	11.0	0.2	Rg	0.2	7.9	Rg	1.9	11.9	Rg	32.9	3.3	—	Rg	Rg	Rg	Rg	0.5	—	—
36	Rg	7.0	1.7	25.7	1.1	0.1	2.7	—	0.7	12.0	Rg	Rg	Rg	12.4	Rg	2.0	12.9	1.2	20.9	5.0	—	Rg	0.3	—	Rg	Rg	—	—
37	Rg	6.4	3.9	26.2	2.2	0.1	3.4	Rg	1.3	13.6	Rg	Rg	Rg	2.2	Rg	1.3	8.1	2.7	18.2	2.3	—	Rg	0.3	Rg	Rg	0.3	—	—
48	—	0.4	11.3	6.7	1.4	Rg	3.2	Rg	Rg	10.1	Rg	—	0.5	6.2	Rg	9.3	32.4	—	13.2	8.0	0.2	0.7	Rg	—	0.6	0.1	—	—
49	I	2.9	5.3	16.6	1.8	0.1	3.7	Rg	0.8	13.0	Rg	0.1	Rg	1.7	0.6	4.6	18.3	0.9	12.7	8.1	0.3	0.3	Rg	Rg	Rg	0.1	Rg	—
50	Rg	3.3	4.0	13.2	0.6	Rg	2.5	—	Rg	11.0	0.4	0.1	—	1.6	Rg	0.5	29.6	0.8	20.1	7.0	Rg	—	0.3	—	0.3	0.1	—	—
51	Rg	5.1	2.2	18.7	0.1	Rg	1.6	Rg	Rg	13.8	0.2	—	Rg	1.9	0.3	2.2	23.6	Rg	22.2	5.6	Rg	Rg	0.3	Rg	0.1	Rg	—	—
52	Rg	0.8	6.0	17.7	0.2	Rg	—	Rg	1.2	23.3	0.1	Rg	—	0.9	1.0	5.5	20.8	0.4	24.2	4.3	Rg	—	Rg	—	0.1	Rg	—	—
53	Rg	0.2	5.2	7.9	1.2	Rg	2.4	Rg	Rg	10.1	0.2	—	Rg	0.8	0.3	3.3	44.2	Rg	13.7	6.6	Rg	—	0.5	Rg	0.6	Rg	—	—
54	Rg	1.5	—	8.6	0.4	Rg	Rg	Rg	Rg	13.5	0.1	Rg	—	1.0	0.1	3.3	35.5	Rg	21.5	7.9	Rg	Rg	0.3	0.6	Rg	Rg	—	—
55	Rg	1.1	12.6	2.4	0.4	Rg	1.6	Rg	0.2	8.3	0.1	Rg	—	1.2	0.1	5.2	40.6	Rg	14.3	10.5	Rg	0.2	Rg	Rg	—	Rg	—	—

130

TABLE 50 (cont'd). Mineralogical Composition (%) of Heavy Residue of Sands of Northwest Caspian Region

No. of samples	Corundum	Hematite	Limonite	Ilmenite	Rutile	Anatase	Leucoxene	Spinel	Chromite	Magnetite	Sillimanite	Kyanite	Staurolite	Zircon	Anthophyllite	Amphiboles	Orthorhombic pyroxenes	Monoclinic pyroxenes	Garnets	Epidote	Olivine	Tourmaline	Chlorite	Sphene	Monazite	Apatite	Carbonates	Barite
1	2	3	4	5	6	7	8	9	10	11	12	13	14	15	16	17	18	19	20	21	22	23	24	25	26	27	28	29
56	Rg	1.9	9.3	6.5	0.4	Rg	0.4	Rg	0.6	11.9	0.1	Rg	—	1.2	0.4	4.4	32.5	—	26.9	3.5	—	Rg	Rg	Rg	Rg	—	—	—
57	Rg	0.8	6.2	24.4	1.1	Rg	2.2	Rg	0.6	18.2	Rg	2.4	0.6	4.9	—	1.4	17.5	0.6	15.2	5.6	Rg	—	Rg	Rg	—	0.7	—	—
68	Rg	0.3	8.4	17.0	3.3	Rg	0.7	—	1.5	14.0	Rg	—	Rg	7.4	Rg	2.1	8.0	—	27.0	6.9	—	0.6	Rg	Rg	Rg	Rg	—	—
70	Rg	1.5	2.9	21.5	0.5	0.1	1.8	Rg	1.2	15.0	0.1	0.1	0.1	2.7	0.2	1.3	20.4	—	19.3	6.7	—	0.3	Rg	Rg	—	0.1	—	—
71	Rg	8.7	0.7	17.9	1.2	0.1	2.7	Rg	1.0	15.0	Rg	Rg	Rg	4.6	Rg	Rg	—	—	17.0	4.2	—	0.1	—	0.4	Rg	Rg	—	Rg
72	Rg	6.7	17.3	5.0	0.2	0.1	2.2	Rg	0.2	29.0	Rg	Rg	Rg	6.4	—	Rg	19.0	Rg	9.9	1.9	—	0.7	Rg	0.5	0.2	Rg	—	—
73	Rg	Rg	13.3	16.4	0.2	0.2	2.2	Rg	0.4	27.5	Rg	Rg	Rg	3.0	—	1.7	20.9	Rg	12.8	5.3	—	0.8	Rg	Rg	Rg	Rg	—	—
74	Rg	0.3	4.5	17.4	1.0	Rg	1.2	Rg	0.6	22.0	0.9	Rg	Rg	4.0	Rg	1.1	25.4	Rg	14.3	6.2	—	1.2	Rg	Rg	—	Rg	—	Rg
75	Rg	—	Rg	15.4	0.1	0.1	2.4	Rg	0.8	23.1	1.0	0.2	Rg	0.3	Rg	2.8	27.1	—	24.5	1.3	—	Rg	Rg	Rg	—	—	—
76	Rg	Rg	8.8	15.4	0.5	Rg	1.9	Rg	0.4	12.0	1.0	Rg	Rg	0.6	0.4	Rg	25.7	—	18.3	10.6	—	0.6	Rg	Rg	—	0.3	—	—
77	Rg	Rg	2.2	16.8	0.2	Rg	0.5	Rg	0.9	27.4	Rg	Rg	Rg	4.6	Rg	Rg	10.1	Rg	22.5	2.2	—	Rg	Rg	Rg	—	0.1	—	—
78	0.1	Rg	1.9	23.1	0.9	Rg	1.8	Rg	0.3	23.9	0.2	Rg	Rg	1.5	Rg	1.5	16.6	Rg	26.9	2.2	—	0.4	Rg	Rg	—	Rg	—	—
79	Rg	Rg	1.5	23.8	1.4	0.1	3.8	Rg	1.8	21.7	0.1	Rg	—	3.8	Rg	1.2	17.5	—	20.3	3.0	Rg	Rg	Rg	Rg	Rg	0.1	—	—
Volga type																												
17	—	12.0	Rg	40.0	4.6	0.3	2.0	Rg	4.9	7.6	—	0.5	Rg	9.7	—	Rg	Rg	Rg	4.3	10.0	—	0.3	Rg	Rg	Rg	Rg	—	—
18	—	10.4	Rg	51.5	5.2	0.3	1.6	Rg	0.3	10.0	—	0.1	1.9	9.6	—	Rg	Rg	Rg	6.9	1.6	—	Rg	Rg	0.6	Rg	Rg	—	—
19	—	2.0	—	61.7	5.4	Rg	1.0	Rg	1.3	7.7	—	Rg	Rg	15.0	—	1.0	Rg	Rg	2.9	2.0	—	Rg	Rg	Rg	Rg	Rg	Rg	—
21	—	6.6	Rg	55.5	4.1	Rg	1.2	Rg	0.9	4.0	—	0.9	0.6	13.6	—	0.3	0.3	Rg	7.3	4.7	—	Rg	Rg	Rg	Rg	Rg	—	—
24	—	4.2	0.3	53.0	5.5	Rg	0.9	Rg	1.3	8.3	—	0.5	0.5	18.4	—	Rg	Rg	Rg	2.6	4.5	—	Rg	Rg	Rg	0.3	0.4	—	—
25	—	3.7	Rg	52.8	5.9	Rg	Rg	Rg	0.5	10.5	—	0.6	1.1	6.9	—	0.6	2.0	Rg	9.8	7.4	—	Rg	Rg	Rg	Rg	0.1	—	—
26	—	3.9	Rg	49.0	4.2	Rg	Rg	Rg	0.3	9.1	—	1.5	1.3	6.9	—	1.3	0.3	Rg	10.5	9.4	Rg	Rg	Rg	Rg	Rg	0.4	—	—
40	—	11.3	0.3	43.1	4.0	0.1	2.4	Rg	1.7	4.9	—	0.5	1.0	10.4	—	3.9	1.5	—	14.3	4.0	—	0.3	Rg	1.3	0.3	0.3	—	—
43	—	11.9	1.2	23.8	1.5	Rg	3.1	Rg	2.7	7.6	—	0.2	0.6	6.4	—	0.3	2.1	Rg	10.7	19.9	—	Rg	Rg	0.3	0.2	0.3	—	—
61	—	3.8	3.8	34.5	4.7	Rg	1.6	Rg	3.1	15.4	—	0.7	1.0	9.1	—	0.8	2.7	0.2	10.2	8.2	—	Rg	Rg	0.3	Rg	0.1	—	—
62	—	6.3	Rg	37.9	5.2	Rg	5.7	Rg	5.1	12.0	Rg	0.2	0.8	6.2	—	0.8	2.7	Rg	14.7	4.3	—	1.0	Rg	0.5	0.2	0.1	—	—
63	—	7.5	Rg	40.3	3.7	0.1	2.9	Rg	0.6	3.7	—	0.8	1.2	9.7	Rg	Rg	2.7	—	17.1	7.2	—	Rg	Rg	0.7	Rg	Rg	—	—
64	—	3.1	Rg	56.6	4.1	Rg	2.2	—	1.0	5.0	Rg	1.3	1.4	6.2	—	Rg	0.4	—	8.5	2.8	—	0.3	0.2	0.9	0.7	Rg	—	—
65	—	3.5	7.6	41.4	3.1	Rg	3.0	Rg	1.4	12.0	Rg	1.3	0.8	6.2	Rg	0.8	8.1	Rg	7.3	2.7	—	0.4	Rg	—	Rg	Rg	—	—
67	—	5.1	1.5	37.9	4.1	Rg	2.2	—	0.4	10.4	Rg	Rg	1.3	9.8	—	1.5	4.2	—	14.8	6.6	—	Rg	Rg	p.3.	—	Rg	—	—
80	Rg	—	Rg	41.0	6.8	0.1	1.6	—	7.7	6.4	0.2	0.1	0.8	10.3	—	1.2	Rg	—	14.7	8.2	Rg	Rg	Rg	p.3. 0.4	0.4	0.1	—	—

131

TABLE 50 (cont'd). Mineralogical Composition (%) of Heavy Residue of Sands of Northwest Caspian Region

Mixed type

No. of samples	Corundum	Hematite	Limonite	Ilmenite	Rutile	Anatase	Leucoxene	Spinel	Chromite	Magnetite	Sillimanite	Kyanite	Staurolite	Zircon	Antho-phyllite	Amphiboles	Orthorhombic pyroxenes	Monoclinic pyroxenes	Garnets	Epidote	Olivine	Tourmaline	Chlorite	Sphene	Monazite	Apatite	Carbonates	Barite
7	—	3.7	6.9	29.0	4.2	0.5	4.2	—	3.7	33.3	—	Rg	Rg	Rg	Rg	0.8	3.4	—	6.1	3.9	—	Rg	—	0.3	—	Rg	—	—
8	—	3.1	1.7	20.6	0.9	Rg	0.6	Rg	Rg	38.5	—	Rg	0.2	4.0	—	Rg	8.0	—	21.3	0.9	—	0.2	Rg	—	—	Rg	—	—
9	—	1.3	2.4	5.4	0.2	Rg	1.7	Rg	0.6	22.4	—	Rg	Rg	2.8	Rg	3.7	33.6	—	12.8	13.1	—	Rg	Rg	—	—	Rg	—	—
11	—	1.7	6.3	6.9	0.3	Rg	1.2	Rg	0.9	25.0	—	Rg	0.3	Rg	3.7	1.2	30.9	0.9	4.0	16.7	—	—	Rg	—	Rg	0.1	—	—
12	—	1.8	0.7	18.4	0.7	0.1	0.5	—	2.4	60.0	—	Rg	0.2	Rg	Rg	0.4	7.8	—	5.7	1.3	—	Rg	Rg	—	—	Rg	—	—
13	—	3.0	0.2	18.5	1.9	1.4	2.0	—	0.5	Rg	Rg	0.4	Rg	3.2	—	0.7	11.8	0.7	12.6	4.6	—	Rg	Rg	0.3	—	Rg	—	—
14	—	4.5	Rg	52.0	5.9	Rg	5.9	Rg	2.0	2.0	Rg	Rg	0.7	9.7	—	0.3	Rg	Rg	4.8	12.1	—	Rg	Rg	0.3	0.3	Rg	—	—
15	—	2.0	0.6	50.0	5.0	0.3	4.2	Rg	3.7	4.0	—	0.2	0.3	8.1	—	Rg	0.3	0.3	9.6	11.1	—	0.3	Rg	Rg	0.3	Rg	—	—
16	—	14.5	—	43.5	4.0	0.1	2.5	—	3.1	16.7	—	0.6	0.6	4.0	—	Rg	Rg	Rg	3.7	5.9	—	Rg	Rg	0.5	Rg	Rg	—	—
22	—	3.2	—	53.8	6.7	0.2	0.6	Rg	1.5	7.4	—	Rg	0.9	16.4	—	Rg	Rg	Rg	5.9	3.2	—	0.3	Rg	0.6	Rg	Rg	—	—
23	—	1.4	—	32.5	1.5	Rg	Rg	Rg	2.7	11.2	Rg	Rg	0.5	9.3	—	Rg	1.2	Rg	2.4	1.0	Rg	—	—	—	—	Rg	—	—
27	—	6.4	2.0	44.3	5.2	Rg	0.1	—	0.8	18.8	—	0.3	2.2	8.2	—	0.4	6.3	0.2	2.2	4.6	—	0.3	0.3	Rg	0.3	1.1	—	—
28	—	2.7	2.6	35.2	2.4	Rg	15.2	—	2.7	23.1	—	Rg	0.6	6.0	Pg	3.2	9.2	1.2	13.4	2.4	—	—	Rg	0.6	Rg	Rg	—	—
38	—	6.3	0.6	20.2	0.8	0.1	1.3	—	0.8	7.1	0.1	0.7	—	13.3	—	1.1	0.8	0.3	17.8	8.5	—	0.3	0.3	Rg	Rg	Rg	—	—
39	—	10.4	1.1	36.7	3.1	Rg	2.2	—	2.6	8.3	0.1	0.6	0.6	9.4	—	2.6	1.0	0.4	17.9	2.4	—	—	Rg	0.3	Rg	Rg	—	—
41	—	25.2	Rg	42.2	4.1	Rg	2.9	Rg	3.9	6.2	Rg	0.8	0.6	11.5	—	1.8	Rg	Rg	12.5	3.9	—	0.4	Rg	0.1	Rg	Rg	—	—
42	—	15.4	Rg	29.3	4.6	Rg	0.8	Rg	6.2	3.3	Rg	0.6	0.4	11.1	—	0.9	1.0	0.4	11.7	10.6	—	0.3	Rg	0.6	Rg	0.1	—	—
44	—	14.2	1.0	29.2	3.2	0.1	0.4	Rg	0.5	8.7	Rg	1.7	0.8	11.1	—	Rg	0.8	1.0	12.9	11.3	—	Rg	Rg	Rg	Rg	Rg	—	—
45	—	17.9	Rg	35.2	6.0	0.3	1.2	Rg	1.6	8.7	Rg	0.6	0.3	12.7	—	Rg	0.3	Rg	10.9	5.1	—	Rg	Rg	0.6	Rg	Rg	—	—
46	—	Rg	Rg	42.2	3.8	0.1	5.4	Rg	4.1	8.3	Rg	1.3	1.8	6.3	—	0.9	6.9	0.2	15.7	12.8	—	0.9	Rg	0.2	Rg	0.1	—	—
47	Rg	8.7	1.2	25.1	1.9	0.1	0.1	—	0.5	11.1	Rg	0.4	Rg	12.0	—	1.7	6.0	Rg	20.1	5.2	—	Rg	Rg	0.2	0.3	Rg	—	—
58	—	2.7	3.8	32.9	1.9	Rg	0.1	—	1.1	22.2	0.1	Rg	0.6	11.4	—	0.3	18.6	0.2	15.1	2.2	—	Rg	Rg	0.2	0.3	0.1	0.3	—
59	—	0.9	19.3	12.7	0.9	Rg	—	—	0.9	12.5	0.1	0.6	0.2	3.3	—	5.4	8.7	2.8	11.2	9.7	—	Rg	Rg	0.2	0.2	Rg	—	—
60	Rg	3.4	4.3	32.3	2.1	Rg	0.6	Rg	0.5	16.0	0.1	0.6	1.3	3.2	—	1.4	3.1	0.7	18.1	8.4	—	Rg	Rg	—	0.2	0.1	—	—
66	—	9.8	1.3	33.6	4.7	0.1	0.8	—	0.5	6.1	0.1	0.1	1.3	5.0	—	3.0	—	—	21.2	7.7	—	Rg	Rg	0.6	Rg	Rg	—	—
69	—	4.3	1.9	31.2	2.3	0.1	0.8	—	3.0	24.0	p. 3.	—	0.3	5.6	—	1.8	6.5	Rg	13.9	4.3	—	Rg	Rg	—	Rg	Rg	—	—

Rutile – rounded circular grains, occasionally prismatic, bright red in color; black nigrins occur.

Spinel – rounded grains or angular fragments of a bluish or bluish-green color.

Andalusite – angular fragments, pale pink, pleochroic.

Zircon – mainly in the form of rounded grains of pale pink color with a very small amount in the form of prismatic crystals of the hyacinth or zirconite type, colorless or with a violet tinge.

Amphiboles – elongate prismatic grains of hornblende of a dark green, almost black color with Ng = 1.694, Np = 1.666, angle of cleavage c∧Ng = 16°, or columnar elongate grains of basaltic hornblende of a dark brown color with the small extinction angles characteristic of basaltic hornblendes.

Orthorhombic pyroxenes – long prismatic grains of green color, pleochroic from pink to green, Ng = 1.694, Np = 1.682; or greenish brown with higher refractive indices: Ng = 1.706, Np = 1.696. Thus, the orthorhombic pyroxenes found are hypersthenes with different Mg : Fe ratios.

Monoclinic pyroxenes of the diopside-hedenbergite series – angular grains of irregular shape, some bluish, with Ng = 1.730-1.734, Np = 1.706-1.714, extinction angles of about c∧Ng = 45°.

Garnets – angular fragments of pale pink color, some containing opaque point inclusions.

As a whole, the fragmentary material of sands of the Sulak type is large-grained with various degrees of rounding – angular or half-rounded rock fragments predominate. The predominate mineral in the heavy fraction is limonite and sometimes ilmenite or zircon.

Sands of the Terek Type

The sands lying along the lower course of the Terek River and occupying the central and southwestern part of the experimental polygon consist of typical graywacke varieties with poorly sorted, fairly large-grained fragmentary material. They consist mainly of fragments of effusive rocks, pyroxene andesites, porphyries, and many fragments of granites. The andesite fragments consist of black compact augite and hypersthene andesites with a characteristic pilotaxitic structure. The phenocrysts are large and consist predominantly of orthorhombic pyroxenes. Fragments of plagioclase andesites with a groundmass of the same pilotaxitic structure occur. The groundmass is usually chloritized. Acid representatives of effusive rocks – quartz porphyries and their tuffs – are occasionally found in the fragments. Fragments of coarse-grained biotite granites are fairly common in sands of the Terek type, but their disintegration is usually more advanced so that the quartz or feldspar grains are included in the grain count and, hence, the granites are not included in the category of rock fragments. In addition, fragments of cataclastic fine-grained greisenized granites

Fragments of metamorphosed siltstones, silty pelites, and clayey shales are present in all samples of sands of the Terek type. Fragments of limestones, similar in habit to those in sands of the Sulak type, are present here in much smaller amounts (see Table 49).

The feldspar fragments consist of relatively fresh acid plagioclases and more basic varieties with broad polysynthetic twins. Zonal plagioclases are fairly frequent. Predominantly potassic feldspars are usually represented by microcline with gridiron twinning, sometimes exhibiting perthitic structure, with the albitic ingrowths mottled or banded. Biotite and pyroxenes, mainly orthorhombic (hypersthenes), predominate among the fragments of dark-colored minerals; crystals of green hornblende and brownish basaltic hornblende are present.

The heavy residues washed from sands of the Terek type are coarse-grained. The grain size varies from 0.05 to 0.7 mm, with grains of 0.24-0.50 mm predominating. The freshness of the material, the well-preserved nature of the crystalline forms, and the predominance of prismatic grains are characteristic. The morphological characteristics of some of the minerals of the heavy fraction are given below.

Corundum – angular fragments of mottled color (blue spots in transparent grains), sometimes in the form of flat bluish grains often containing inclusions.

Ilmenite – large tabular crystals, occasional aggregates, moderately rounded, with shining faces, and sometimes with accumulations of leucoxene in pits.

Rutile – prismatic grains, rounded with flat ends, and bright red. Black varieties (nigrines) with distinct adamantine luster occur.

Spinel – angular or rounded bluish-green fragments.

Chromite – octahedra with slightly blunted edges, black, mat.

Magnetite – octahedra, but often with unevenly developed faces. The crystals have blunt apices and are black with a pronounced luster. Individual grains of magnetite occur in the form of inclusions within crystals of orthorhombic pyroxenes.

Kyanite – rounded transparent colorless grains of rare occurrence. The kyanite content of the heavy residue increases toward the north at the expense of sillimanite, which is a little better represented in the southwest. Sillimanite consists of long transparent columnar crystals. Grains of pink andalusite occur.

Zircon – prismatic crystals of dipyramidal habit, slightly blunted at the apices. The crystals are pale pink or transparent and almost colorless.

Amphiboles (ordinary hornblende, basaltic hornblende, very rarely actinolite and kaersutite) – in the form of long prismatic grains, half-rounded or with blunt corners.

Orthorhombic pyroxenes, which constitute the main volume of the elctromagnetic fraction, consist of semitransparent pale green prismatic grains, which often contain inclusions of ore minerals. In view of the fact that some orthorhombic pyroxenes are attracted into the magnetic fraction, these inclusions are magnetite grains. The refractive indices are Ng = 1.710, Np = 1.702. Strong pleochroism in greenish and pink tones, very characteristic of hypersthene, is observed.

Monocline pyroxenes are represented by prismatic grains of the diopside-hedenbergite series. The color is green, the extinction angles large, and the refractive indices are Ng = 1.718, Np = 1.694. Manganese varieties (piedmontite) of a dark brown to black color and with strong pleochroism (Ng = 1.782, Np = 1.767) occur. Blue-green grains of aegirine occasionally occur.

Olivine – rounded grains, yellowish or colorless with Ng = 1.698, Np = 1.662 or with Ng = 1.706, Np = 1.670.

Garnets – in the form of sharp-angled fragments of irregular shape. Pale pink varieties and, in several cases, bright green uvarovites predominate.

Epidote is a relatively rare mineral in the heavy residue of sands of the Terek type. Its grains are yellowish-green, transparent, with smoothed outlines.

Tourmaline – in the form of rounded grains with strong pleochroism from a golden brown to a dark, almost black color.

Apatite – clean transparent or slightly whitish grains, relatively large, well rounded.

As a whole, the fragmentary material of sands of the Terek type consists of large, poorly sorted fragments. The grains are semiangular, mainly prismatic. The abundance of fragments of augite and hypersthene andesites and of coarse-grained granites indicates the proximity of a supply source. The similarity of the rock fragments to the lavas of the Kel'skoe plateau and the granites of the Dar'yal'skii massif determines the location of the supply source. The fragments of acid effusive rocks are most probably due to a supply of material from the right tributary of the Malka River from the Nal'chik region, where they have been described by Rengarten (1955).

Sands of the Volga Type

The sands occurring in the north of the polygon are represented by subgraywackes containing 65-70% quartz. The fragmentary material of the sands is well rounded and better sorted than the material of the Terek sands.

The fragmentary material always contains, in addition to ordinary quartz and feldspars, a certain number of fragments of granites, quartz porphyries, shales, and limestones (see Table 49).

Quartz occurs in the form of well-rounded, clean, transparent grains. The predominantly potassic feldspars also consist of well-rounded transparent or semitransparent grains of microclines, in many cases albitized, with fine perthitic intergrowths.

The rock fragments include (in diminishing amounts) quartz porphyries, siltstones, hornfelses, chert, fine-grained granites, and limestones. The grains of minerals in the heavy fraction have a distinct rounded equant form. The grain size is 0.05-0.25 mm with grains of 0.1-0.15 mm predominating. The heavy residues are very rich in ilmenite, the amount of which reaches 62% with $\overline{X}$ = 45% and S = 9.5% (see Table 50).

The following minerals of the heavy fraction are characteristic of sands of the Volga type:

Hematite – rounded grains or earthy formations, dark brown or almost black.

Ilmenite – grains lamellar, but of a more or less equant shape. The surface of the faces is shiny, uneven, covered with little pits. The degree of rounding is high.

Apatite – in the form of transparent or semitransparent, sometimes whitish, round grains; their surface is also finely pitted.

Rutile – long grains, well rounded, bright red, dark cherry red, or black (nigrin). The faces of the rutile grains are smooth, striations are never found, and even a shagreen surface is sometimes observed. The luster is strong and adamantine. The red varieties are quite transparent.

Leucoxene – earthy aggregates, pale yellow or dirty greenish gray, relatively rare.

Spinel – rare grains, smooth, well rounded, bluish.

Chromite and chrompicotite – octahedral grains, relatively weakly rounded.

Magnetite – in the form of grains of almost equant form, 0.1 mm in size, completely rounded, black, and shining.

Kyanite – lamellar grains with blunt and smoothed corners. The grains are grayish owing to inclusions or are colorless and transparent.

Staurolite – large orange-yellow grains, with a greasy luster, lenticular and round.

Zircon – pale pink transparent grains, completely rounded, some almost spherical in form. Long crystals of colorless zircon, but also with blunt corners, occasionally occur.

Amphiboles are represented by rounded elongate grains of green hornblende, $Ng = 1.706$, $Np = 1.702$, $c \wedge Ng = 12-14°$; brownish-yellow grains of basaltic hornblende with $Np = 1.702$ and $Ng = 1.726$, $c \wedge Ng = 5°$, and bluish-green, distinctly rounded grains of actinolite.

Orthorhombic pyroxene (hypersthene) – rounded grains of prismatic form, transparent, surface of crystals weathered with erosion or solution pits, $Np = 1.690$, $Ng = 1.698$ or $Np' = 1.702$, $Ng = 1.705$, pleochroism in greenish-pink hues.

Monoclinic pyroxenes – in the form of green diopside grains with inclusions of ore minerals, $Np = 1.699$, $Ng = 1.721$, $c \wedge Ng = 43°$, or in the form of brownish-green grains of augite with characteristic extinction in the form of an hour glass.

Aegirines – flat prismatic fragments, rounded, dense green, pleochroic to greenish blue, $Np = 1.754$, $Ng = 1.780$.

Piedmontite – in the form of small, elongate brownish-black grains.

Garnets consist of rounded yellow or brown grains, more angular pink fragments of almandite, and in a few cases bright green uvarovites are present.

Epidote – equant smoothed grains with a shagreen surface and of a yellowish-green color.

Tourmalines in heavy residues from Volga sands vary in color: grayish green with $Np = 1.622$ $Ng = 1.655$, pleochroic from pale pink to brownish green. Rounded golden-brown grains showing pleochroism from pinkish yellow to almost black, with $Np = 1.633$, $Ng = 1.660$, are much rarer.

Sphene – rounded flat grains, yellowish, semitranslucent, with a slightly pitted surface and strong luster.

Some of the heavy residues contain accumulations of carbonates and sulfates. For instance, in heavy residues from some parts of the dry bed of the Kuma River the celestite content reaches 40% of the total heavy fraction.

Thus, petrographic-mineralogical analyses of the investigated sands in the experimental polygon showed that three petrographic types could be clearly distinguished:

1. Sands of the Sulak type – graywackes, consisting mainly of fragments of sedimentary rocks. The heavy fraction of these sands is very rich in limonite and ilmenite.

2. Sands of the Terek type – graywackes, consisting mainly of fragments of effusive rocks (pyroxene andesites). Pyroxenes and magnetite predominate in the heavy residues of these sands and the garnet content is high.

3. Sands of the Volga type – subgraywackes, approximating to monomictic quartz sands, with well-sorted and rounded fragmentary material. Ilmenite predominates in the heavy fraction of these sands and the zircon content is high.

The results of spectrophotometry of the above-described types of sands are given in the form of the polynomial coefficients of the ρ_λ curves of these sands in Tables 51 and 52.

Table 51 shows that the values of ρ_λ for sands of the different petrographic types are different.

Sands of the Sulak type have low values of ρ_λ; the maximum values do not exceed 20%. The slope of the ρ_λ curves is very gentle and they are almost parallel to the wavelength axis (see Fig. 63).

Sands of the Terek type have ρ_λ values in the range 9 to 30%. The slope of the ρ_λ curves is gentle but, as distinct from the ρ_λ curves characteristic of Sulak sands, the curves show small maxima in the wavelength range 580-620 mμ (see Fig. 63).

Sands of the Volga type have high values of ρ_λ, up to 43%, and the ρ_λ curves have a very characteristic shape with a sharp maximum in the wavelength range 600-620 mμ (see Fig. 63).

The values of the polynomial coefficients of the ρ_λ curves of the investigated sands (see Table 52) were compared for two petrographic types – the Volga and Terek types – since the number of observations for sands of the Sulak type was very small. The results of plotting the frequency distribution of the values of the polynomial coefficients for each of the two named types of sands are given in Fig. 65, which shows that the sands can be distinguished from one another by the values of a_2 and a_3. To evaluate the difference we calculated the contingency criterion. Fourfold contingency tables for the values of the polynomial coefficients and petrographic types of the sands showed that (Table 53):

TABLE 51. Values of Polynomial Coefficients of ρ_λ Curves of
Sands of Northwest Caspian Region

No. of Sample	a_0	a_1	a_2	a_3
		Sands of Sulak type		
1	90.43455	2.600505	—0.104607	0.001421
2	316.67577	0.197099	0.728553	—0.021248
3	77.94931	3.145288	0.032097	—0.003511
4	42.72349	4.636527	—0.161567	0.001750-
5 a	71.75198	—0.532057	0.233472	—0.007524
5 b	85.98126	—0.535016	0.199914	—0.006586
5 c	92.97250	—0.939809	0.294221	—0.009407
6	128.68140	—0.402487	0.506788	—0.017044
		Sands of Terek type		
10	112.58660	0.856010	0.382382	—0.012883
29	112.43709	0.888656	0.151929	—0.005585
30	135.82317	1.205002	0.055353	—0.003330
31	92.36715	6.205888	—0.282536	0.004182
32	77.20128	6.004529	—0.124559	—0.001045
33	130.44684	2.622359	0.055059	—0.004026
34	45.59058	15.341026	—0.772263	0.011850
35	152.77272	3.503847	0.082630	—0.005721
36	116.67208	6.753993	—0.211246	0.001548
37	69.06940	8.124820	—0.355983	0.004753

TABLE 51 (cont'd). Values of Polynomial Coefficients of ρ_λ Curves
of Sands of Northwest Caspian Region

No. of Sample	a_0	a_1	a_2	a_3
48				
49	85.07039	2.013545	0.020068	—0.002139
50	109.39216	6.245811	—0.317012	0.005704
51	134.22376	6.192143	—0.120655	—0.000903
52	145.81742	6.973545	—0.273273	0.003162
53	122.09402	4.149418	—0.0208100	0.003569
54	131.41070	5.842804	—0.282065	0.004599
55	138.79437	3.151174	0.009321	—0.002187
56	107.59518	2.363492	0.087550	—0.003567
57	81.67732	0.938268	0.080818	—0.002636
68	121.73625	3.738723	—0.098573	0.000080
70	121.35193	2.064803	0.583234	—0.015568
71	112.91860	7.496081	—0.338108	0.005486
72	146.77095	4.166514	0.131423	—0.007467
73	114.53728	8.734838	—0.150048	—0.7001381
74	169.56429	9.717749	0.071914	—0.010172
75	121.10297	6.100648	—0.087096	—0.001596
76	144.63787	8.161764	—0.183657	—0.000621
77	105.72045	4.632436	—0.105374	0.000299
78	296.77224	1.558501	0.383729	—0.013064
	91.20780	1.578799	0.047805	—0.001562

Sands of Volga type

No. of Sample	a_0	a_1	a_2	a_3
17	131.28262	7.433427	0.623127	—0.023806
18	111.18736	—6.757583	—0.631432	—0.023742
19	108.78893	6.798271	0.420437	—0.015942
21	139.91006	9.959188	0.559086	—0.023003
24	73.69688	3.029142	0.037150	—0.001881
25	111.97933	1.349368	0.674584	—0.020321
26	253.08682	7.303342	0.346401	—0.018188
40	129.22136	12.340418	—0.071724	—0.007783
43	121.97650	4.973968	0.564576	—0.021327
61	112.86801	5.433750	0.178766	—0.008882
62	106.93600	6.071935	—0.032634	—0.002891
63	122.45862	9.782979	—0.008389	—0.004468

No. of Sample	a_0	a_1	a_2	a_3
64	177.43463	−7.004228	1.672341	−0.043478
65	241.75087	15.453324	−0.324589	0.000011
67	156.24495	−3.162608	1.716108	−0.048509
80	174.91857	1.547012	1.003835	−0.029337

Sands of mixed composition

No. of Sample	a_0	a_1	a_2	a_3
7	68.15513	0.791555	0.210691	0.007817
8	97.49779	1.234179	0.364132	0.011371
9	92.99415	0.996406	0.368973	0.011754
11	158.13517	4.553113	0.209253	0.002599
12	171.89015	6.014458	0.366544	0.007171
13	51.30168	7.009223	0.277857	0.003000
14	144.22132	4.853709	0.117315	0.001107
15	276.49889	8.913415	0.296015	0.001282
16	223.90028	10.887273	0.554098	0.009718
22	128.21101	1.808360	0.089408	1.004178
23	86.12096	2.156042	0.321824	1.010031
27	101.58091	1.601167	0.179811	0.006636
28	84.10153	2.677827	0.018207	0.001321
38	110.05233	8.314015	0.348913	0.004087
39	143.02278	14.962360	0.785396	0.012031
41	99.03155	13.616050	0.277381	0.001571
42	110.76890	12.594626	0.065895	0.007334
44	127.42573	6.569858	0.384584	0.016843
45	117.72355	1.808135	0.508599	0.016357
46	116.05419	3.184727	0.264907	0.009916
47	89.63829	1.236139	0.104375	0.004212
58	112.91860	7.496081	0.338108	0.005486
59	99.95849	3.213701	0.025670	0.000989
60	93.33208	3.516597	0.047483	0.000651
66	267.11594	18.881408	0.284275	0.003960
79	117.91667	0.048595	0.247973	0.006605

a) sands of the Volga type can be reliably distinguished from Terek sands from the values of a_2 and a_3;

b) sands of the Sulak type are indistinguishable in values of a_2 from Terek and Volga sands; from the values of a_3 they can be separated with some difficulty ($\chi^2 = 4.7$) from Terek sands and a little more reliably ($\chi^2 = 5.2$) from Volga sands.

Thus, the calculated values of χ^2 (Table 53) show the existence of a relationship between the type petrographic composition of sands and the values of the polynomial coefficients a_2 and a_3. From the value of a_2 we can determine if a sand belongs to the Terek or Volga type, and from the value of a_3 we can determine if a sand belongs to one of the three described petrographic types of sands.

An additional check of the discovered association between the ρ_λ values of sand and its petrographic-mineralogical composition was made by calculating the correlation coefficients r between the values of a_2 and a_3, on

TABLE 52. Mean Values of Polynomial Coefficients of ρ_λ Curves of Sands from Northwest Caspian Region

Polynomial coefficients of ρ_λ curves	Type of sand and number of observations					
	Sulak, 8		Terek, 30		Volga, 16	
	$\bar{X}$	S	$\bar{X}$	S	$\bar{X}$	S
a_0	113.40	80.02	121.6	41.84	142.11	47.16
a_1	1.02	1.98	4.72	3.49	4.66	6.13
a_2	0.22	0.29	−0.059	0.260	0.42	0.63
a_3	−0.0078	0.0077	−0.0015	0.0061	−0.018	0.014

TABLE 53. Fourfold Tables of Contingency of Polynomial Coefficients for Sands of Different Type from Northwest Caspian Region

Type of sand	a_2		χ^2	a_3		χ^2
	$\leqslant +0.3$	$>+0.3$		$\leqslant -0.015$	>-0.015	
Terek . . .	30	1	19.2	1	30	22.0
Volga . . .	6	9		10	5	

Type of sand	a_2		χ^2	a_3		χ^2
	$\leqslant 0.0$	>0.0		$\leqslant -0.002$	>-0.002	
Terek . . .	16	14	1.4	12	18	4.7
Sulak . .	2	5		6	1	

Type of sand	a_2		χ^2	a_3		χ^2
	<0.25	>0.25		$\leqslant -0.01$	>-0.01	
Sulak . . .	5	2	1.9	1	6	5.2
Volga . . .	6	9		10	5	

TABLE 54. Correlation Coefficients Between Composition of Sands and Parameters of Polynomial Coefficients a_2 and a_3

Composition of sands	a_2		a_3		No. of observations
	r	t	r	t	
Quartz.	+0.85	4.4	−0.84	4.2	15
Rock fragments	−0.79	3.7	+0.70	3.0	15
Fraction 0.105-0.149 mm	+0.24	2.0	−0.27	2.3	74
Fraction 0.07-0.105 mm	+0.25	2.1	−0.14	1.2	74
Ilmenite	+0.30	2.6	−0.44	4.0	74
Zircon.	+0.17	1.4	−0.29	2.5	74
Orthorhombic pyroxenes	−0.10	0.8	+0.41	3.7	74
Orthorhombic pyroxenes (for Terek sands.	−0.30	1.6	+0.32	1.7	30

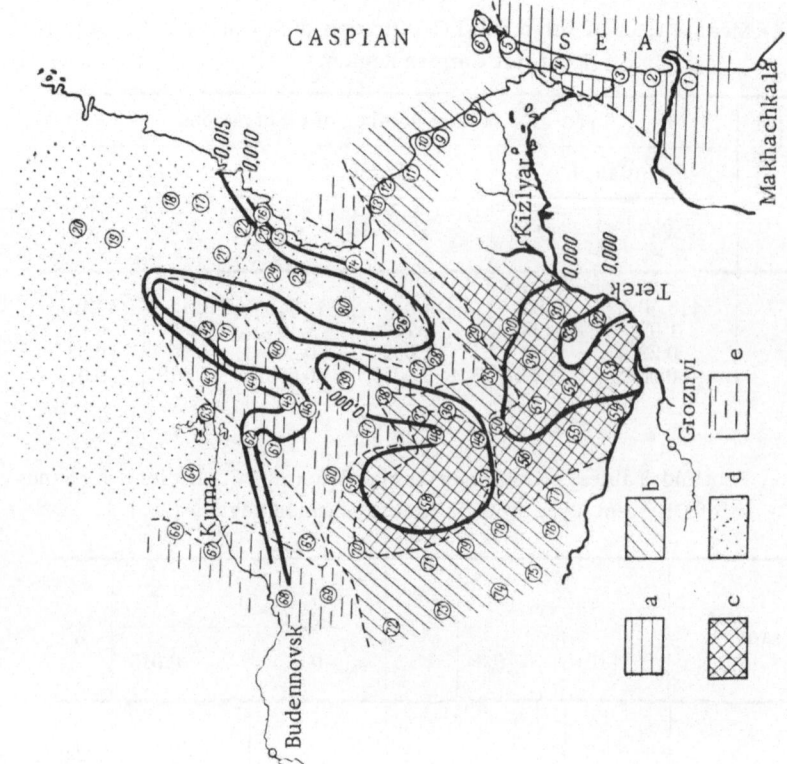

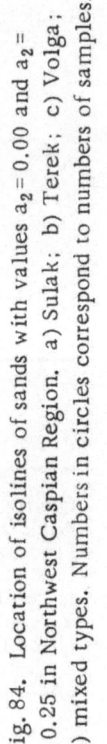

Fig. 85. Location of isolines of sands with values $a_3 = 0.000$, $a_3 = 0.010$, and $a_3 = 0.015$ in the Northwest Caspian Region. a) Sulak; b) Terek; c) Terek with heavy residue containing more than 20% orthorhombic pyroxenes; d) Volga; e) mixed types.

Fig. 84. Location of isolines of sands with values $a_2 = 0.00$ and $a_2 = +0.25$ in Northwest Caspian Region. a) Sulak; b) Terek; c) Volga; d) mixed types. Numbers in circles correspond to numbers of samples.

one hand, and the amount of certain components in the sand samples, on the other. Table 54 gives the values of these correlation coefficients r and their significance t, evaluated from the normalized standard value Z (according to Fisher, see Appendix IV).

An analysis of Table 54 shows that the spectral luminance of the investigated sands is affected by:

a) the content of quartz fragments — a significant positive association with a_2 and an equally strong negative association with a_3;

b) the content of rock fragments (porphyrites and clayey shales) is positively correlated with a_2 and negatively with a_3;

c) the amount of 0.07- to 0.105-mm fraction is positively correlated with a_2; the amount of the 0.105- to 0.149-mm fraction is negatively correlated with a_3;

d) an effect of the content of such minerals as orthorhombic pyroxenes, zircon, and ilmenite on the values of a_1, a_2, and a_3 of the sands can be detected, but these minerals themselves are correlated with the mineral complex of the particular petrographic type. For instance sands of the Volga type are rich in zircon and ilmenite, while sands of the Terek type are rich in orthorhombic pyroxenes.

To compile the trial aeropetrographic maps (diagrams) we used the values of the polynomial coefficients a_2 (Fig. 84) and a_3 (Fig. 85).

In Fig. 84 the isoline corresponding to the value $a_2 = 0.00$ almost coincides with the limit of occurrence of sands of the Terek type, drawn, as already mentioned, through the isoline for 10% content of orthorhombic pyroxenes in the heavy residues from these sands.

The isoline corresponding to the values $a_2 = 0.25$ almost coincides with the limit of occurrence of sands of the Volga type, drawn through the isoline with 40% content of ilmenite in the heavy residues from these sands.

Figure 85 shows the isolines of the values of a_3. The area bounded by the isoline $a_3 = 0.00$ is characterized by sand outcrops containing not less than 20% orthorhombic pyroxenes in the heavy fraction.

The isoline $a_3 = 0.015$ demarcates the area of sands of the Volga type, in which the zircon content of the heavy fraction is 8-10%.

Thus, the results of the experimental aeropetrographic survey of modern sands, carried out in 1957, showed that petrographic mapping of sands can be conducted by measurement of their spectral luminance from the air.

In area mapping of sand outcrops, the measured values of the spectral luminance factors can be converted to the polynomial coefficients of the curves. In this case the values of a_2 and a_3 are used.

Thus, measurement of the spectral luminance of sandy-slit deposits from the air showed that petrographic mapping can be carried out from the air by means of spectrometers and an appropriate treatment of the obtained ρ_λ curves.

SUMMARY

1. The method of determining the composition of rocks from the air by measurement of their spectral luminance factors owes its origin to the progress of science in the field of optics and electronics.

2. The spectral luminance factors of rock outcrops can be measured from the air with the aid of spectrometers operating on the principle of photographic photometry or photoelectric registration of the reflected radiant energy.

3. The observed values of the spectral luminance factors of the investigated rocks must be subjected to mathematical treatment to obtain a quantitative evaluation of the forms of the ρ_λ curves.

4. Correlations between the obtained photometric parameters and the constituents of the rock are revealed by comparison of the data with the aid of the contingency criterion (χ^2 method) or by correlation analysis.

The existence of an association between the values of the photometric parameters of rocks and their lithofacies characteristics is demonstrated by the example of an investigation of West Turkmenian sandy-silt deposits of different age.

5. Aerial spectrophotometry of geological objects and a comparison of the obtained data with the composition of the rocks in the investigated region provides a means of aeropetrographic mapping.

The method has now been developed to the stage of aeropetrographic mapping of sandy-silt deposits and is a most effective method of geological research in such inaccessible regions as sandy deserts.

6. The further development of the method will probably be directed toward a study of the reflecting properties of rocks of other types – igneous, metamorphic, and others. This can be accomplished by the construction of more advanced spectrometers and the wide use of statistical techniques in the geological interpretation of the photometric properties of the rocks.

LITERATURE CITED

Abishev, Kh. A. A determination of the absolute albedo of a magnesia surface. Izvest. Akad. Nauk Kazakh. SSR, Ser. Astrofiz., No. 54, issue 3, 104, 1948.

Abishev, Kh. A. An Investigation of the Reflectance of Mat Surfaces. Alma-Ata, 1949.

Alekseev, S.S. Color Science. Iskusstvo, 1952.

Alekseev, V.A., and Belov, S.V. Spectral reflectance of trees and other objects in an aerial photographic survey of West Ukraine. Aeromethods. Trudy Lab. Aerometodov Akad. Nauk. SSSR, X, 105, 1960.

Ångström, A. A new instrument for measuring sky radiation. Monthly Weather Rev. 4, 1919.

Ångström, A. The albedo of various surfaces of ground. Geograf. Annal., 4, 1925.

Artsybashev, E.S., and Belov, S.V. Reflectance of Woody Materials. Trudy Lab. Aerometodov Akad. Nauk SSSR, 6, 1957.

Artsybashev, E.S. The Spectral Reflectance of Woody Vegetation in Relation to the Interpretation of Aerial Forest Photographs. Leningrad, 1959.

Baranova, Z.E. Stratigraphy, lithofacies characteristics, and oil-bearing prospects of Jurassic deposits of West Turkmenia. Trudy Vsesoyuzni Neftyanoi Nauch.-Issledovatel. Geol.-Razved. Inst., No. 17, 1956.

Belonogova, I.N., and Tolchel'nikov, Yu. S. Relationship between spectral luminance of minerals and degree of dispersion. Izvest. Akad. Nauk SSSR, Ser. Geol., 11, 98, 1959.

Belov, S.V. Aerial Photography of Forests. Leningrad, 1959.

Belov, S.V., and Berezin, A.M. The importance of the conditions of aerial photography and various types of films for forestry research. Trudy Lab. Aerometodov Akad. Nauk SSSR, VI, p. 146, 1958.

Benford, I., Lloyd, G.P., and Schwarz S. Coefficients of reflection of MgO and $MgCO_3$, J. Opt. Soc. Am., 38, 8, 1948.

Berkovich, S.L., Gofman, M.V., Lobachev, M.V., Fal'k, T.K., and Sharonov, D.I. A high-transmission spectrometer with DFS-12 diffraction gratings. Optika i Spektroskopiya, VI, 6, 824, 1959.

Blyumberg, I.B. Technology of Cinefilm Processing. Iskusstvo, Moscow, 1958.

Boiko, P.N. An automatically recording spectroelectrophotometer. Izvest. Aerofiz. Inst. Akad. Nauk Kazakh.SSR, VIII, p. 108, 1959.

Boiko, P.N. Livshits, G. Sh., and Toropov, T.P. Projector investigations of indicatrices of diffusion. Izvest. Astrofiz. Inst. Akad. Nauk Kazakh.SSR, VIII, p. 98, 1959.

Brustin, B.N., Gilev, S.S., Semenovich, V.S., and Yutskevich, Yu. K. Photographic Equipment for Airplanes. Leningrad, 1958.

Connady, A.E. Applied Optics and Optical Design. New York, 1957.

Danchev, V.I. On the interpretation of the data of lithological investigations of sedimentary rocks. Izvest. Akad. Nauk SSSR, Ser. Geol., No. 5, 1946.

Danchev, V.I. The importance of a numerical characterization of the color of sedimentary rocks. Sovet. Geol., No. 18, 164, 1947.

Danchev, V.I. An experimental lithological investigation of the lower part of the deposits of the Tatarian stage in the Kazan Volga Region. Trudy Inst. Geol. Nauk Akad. Nauk SSSR, No. 87, 1947.

Danchev, V.I. Color of sedimentary rocks as one of the indices of the conditions of their formation. Questions of the Mineralogy of Sedimentary Formations, Bks. 3-4. L'vov State University, 1956, p. 51.

Fabry, G. Introduction to Photometry [Russian translation]. Ob'edinenie Nauch-Tekh. Izd., 1934.

Ford, R.K. Rock colors. Bull. Am. Assoc. Petrol. Geologists, 28, 1944.

Fresnel, A. Selected Works on Optics [translated from French by Z.A. Tsetman]. G.S. Landsberg, ed. Gostekhizdat, No. 7, 1955.

Geological Dictionary. Gosudarst. Nauch.-Tekh. Izd., 1955.

Geology of the USSR, Vol. XXII, Turkmenian SSR, Pt. 1. Gosudarst. Nauch.-Tekh. Izd., Min-vo Geol. i Okhrany Nedr, Moscow., 1957.

Girin, O.P., and Stepanov, B.I. Reflection spectra of colored scattering objects. Zhur. Eksp. i Teoret. Fiz., 27, 4 (10), 467, 1954.

Godin, Yu. N. Basic features of the regional tectonics of Turkmenia from the data of geophysical investigations. Izvest. Akad. Nauk SSSR, Ser. Fiz.-Tekh., Khim. i Geol. Nauk, 4, 15, 1960.

Goldman, M.I., and Merwin, H.E. Color chart for field description of sedimentary rocks. Nat. Research Council, 1928.

Gorokhovskii, Yu. N., and Barteneva, O.D. Atlas of Spectral Photographic Materials. Leningrad, 1942.

Gurevich, M.M. Color and Its Measurement. Izd. Akad. Nauk SSSR, 1950.

Handbook of Cinematographic Technique. 1959.

Harrison, D. Practical Spectroscopy [Russian translation]. IL, Moscow, 1950.

Hulst H. C. van de. Light Scattering by Small Particles. New York, John Wiley, 1957.

Istomin, G.A. Resolution of a photographic system at low contrast values. Doklady Akad. Nauk SSSR, 85, 1001, 1952.

Ivanov, A.P., and Toporets, A.S. A spectrophotometric investigation of mixtures of powdered materials. Izvest. Akad Nauk SSSR, Ser. Fiz., 21, 11, 1502, 1957.

Kalitin, N.N. A pyranometer for measuring scattered atmospheric radiation. Izvest. Glavnoi Geofiz. Observ., No. 3, 1929.

Kalitin, N.N. Technique of measuring reflection and transmission of solar radiant energy by plant leaves. Zhur. Russ. Botan. Obshchestvo, 16, No. 1, 1931.

Kalitin, N.N. The pigmentometer and measurements of skin pigmentation. Voprosy Kurortologii, No. 3, 1937.

Kalitin, N.N. Spectral analysis of the ground surface on the southern coast of the Crimea. Voprosy Kurortologii, Nos. 3-4, 76, 1938.

Kol'tsov, V.V. A cathode-ray spectrometer (spectrovisor). Electronic Instruments in the National Economy. Gosenergoizdat, 1959, p. 23.

Kol'tsov, V.V. Measurements of spectral luminance factors outside the laboratory. Svetotekhnika, No. 12, 8, 1960.

Kozlova, K.I. Spectrophotometry of Plants of Different Climatic Zones by Reflected Light. Izd. Akad. Nauk. Kazakh. SSR, Alma-Ata, 1955.

Krasil'shchikov, L.B., and Pyatkovskaya, N.P. Spectral reflection indicatrices of some surfaces in natural light on a cloudy day. Trudy Glavnoi Geofiz. Observ., 68, 132, 1958.

Krasil'shchikov, L.B., Golikova, O.N., and Novosel'tsev, E.P. Photoelectric measurements of relative spectral luminance factors. Trudy Glavnoi Geofiz. Observ., 68, Leningrad, 152, 1958.

Krinov, E.L. Spectral reflectance of some ground formations. Collection on Aerophotometry, No. 2, 1934.

Krinov, E.L. Relationship between spectral luminance factor of natural objects and direction of light. Izvest. Tsentral. Nauch.- Issledovatel. Inst. Geodezii, Aeros. i Kartog., 5-6, 1935.

Krinov, E.L. Spectral Reflectance of Natural Formations. Izd. Akad. Nauk SSSR, 1947.

Krupp, N. Ya., and Lerner, L.A. The MF-2 Nonrecording Photoelectric Microphotometer. 1950.

Krymgol'ts, G. Ya. On the age of the most ancient deposits on Bol'shoi Balkhan. Doklady Akad. Nauk SSSR, XXV, 1, 101, 1950.

Kulebakin, V.S. On the reflection of light from ground coverings. Trudy Gosudarst. Eksp. Elektrotekhn. Inst., No. 17, 1926.

Kurbatov, V.S. On the age and structure of the core of the Tuarkyr anticline. Trudy Inst. Geol. Akad. Nauk Turkmen. SSR, 1, p. 186, 1956.

Landsberg, G.S. Optics. Gostekhizdat, Moscow, 1957.

Lyalikov, K.S., Belonogova, I.N., Meleshko, K.E., Semenchenko, I.V., and Kharchenko, A.P. A new apparatus and method for the investigation of ground surface spectra. Izvest. Akad. Nauk SSSR, Ser. Fiz., XXIII, No. 10, 1958.

Mashrykov, K.K. Jurassic Carboniferous Deposits of Northwest Turkmenia and Their Position in the Crimea–Caucasus–Caspian Carboniferous Province. Izd. Akad. Nauk. Turkmen. SSR, Ashkhabad, 1958.

Minnaert, M. The Nature of Light and Colour in the Open Air. Dover, 1954. [Russian translation published by Fizmatgiz, 1958.]

Mironova, Z.F., and D'yachkova, T.V. A spectrophotometric method of measuring the albedo of natural underlying surfaces. Vestnik Leningrad Gosudars. Univ., No. 22, issue 4, 89, 1957.

Mitropol'skii, A.K. Technique of Statistical Computations. Fizmatgiz, 1961.

Neporent, B.S. Development of molecular spectroscopy in the USSR in recent years. Uspekhi Fiz. Nauk, XVIII, 1, 13, 1959.

Oliver, R. Quantitive determination of rock colour, Science, 66, 1927.

Orlova, N.S. Determination of the reflection factors in natural conditions at different angles of incidence and reflection. Izvest. Akad. Nauk. Kazakh. SSR, Ser. Astrobotan. issues 1-2, No. 90, 1950.

Orlova, N.S. Theory and practice of the investigation of reflectance with the indicatometer. Uchenye Zapiski No. 273, Leningrad. Gosudarst. Univ., issue 4, 144, 1958.

Ostrovskii, M.A. An investigation of the reflecting properties of asphalt surfacings. Svetotekhnika, 1, p. 11, 1956.

Ostwald, W. Color Science [Russian translation]. Promizdat, Moscow-Leningrad, 1926.

Pannekock, A. Photographische Photometrie der Nordlichen Milchstrasse. Amsterdam, 1933.

Podmoshenskii, I.V., and Shelemina, V.M. On an error in the photometry of spectra. Optiko-Mekh. Prom., 1, 49, 1960.

Popov, O.I. Transmission of lower layers of atmosphere in different parts of the spectrum in the 0.3- to 1-mm region. Optika i Spektroskopiya, 3, 5, 504, 1957.

Prokof'ev, V.K. Photographic Methods of Quantitative Spectral Analysis of Metals and Alloys. Pt. II, Methods. Gostekhizdat, Moscow-Leningrad, 1951.

Radlova, L.N. An investigation of the reflectance of a gypsum screen. Tekh. Fiz., XIII, No. 4-5, 1943.

Rengarten, V.P. On the Tertiary Effusives of the Northern Caucasus. Izvest. Akad. Nauk SSSR, Ser. Geol., No. 4, 1955.

Romanova, M.A. Lithostratigraphy of the upper part of the red beds of one of the structures of West Turkmenia. Trudy Inst. Geol. Akad. Nauk Turkmen. SSR, 1, p. 285, 1956.

Romanova, M.A. Correlation between photometric properties of sandy-silt deposits of red beds of Cheleken Peninsula and their lithological and mineralogical composition. Geology of Transcaspia, No. 1, Lab. Aerometodov Akad. Nauk SSSR, 1958a, p. 5.

Romanova, M.A. Results of an experimental aeropetrographic survey of modern sands of the northwest Caspian Region. Doklady Akad. Nauk SSSR, 120, 3, 625, 1958b.

Romanova, M.A. Problems of the geological interpretation of the photometric properties of rocks (as exemplified by an investigation of sedimentary deposits in West Turkmenia). Trudy Lab. Aerometodov Akad. Nauk SSSR, 8, p. 130, 1959.

Romanova, M.A. Spectral luminance of sands of the southeast Kara-Kum. Doklady Akad. Nauk SSSR, 130, 4, 856, 1960a.

Romanova, M.A. An experimental aeropetrographic survey of modern sands of the northwest Caspian region. Trudy Lab. Aerometodov Akad. Nauk SSSR, IX, p. 40, 1960b.

Romanova, M.A. Spectral brightness of West Turkmenian Jurassic arenaceous rocks and its correlation with rock petrography and sedimentary environment. J. Geol., 69, 3, 1961.

Sanders, C.L., and Middlton, F.E. The absolute spectral diffuse reflectance of magnesium oxide in the near infrared. J. Opt. Soc. Am., 43, 1, 1953.

Sensitometric Manual. Properties of Photographic Materials on a Transparent Backing. Yu. N. Gorkhovskii and S.S. Gilev, ed. Gostekhizdat, Moscow, 1955.

Sharonov, V.V. The universal wedge photometer. Uchenye Zapiskii Leningrad. Gosudarst. Univ., 72, No. 31, 1932.

Sharonov, V.V. Scattering of light by terrestrial formations and aerophotometric methods of investigating it. Collection on Aerophotometry, No. 2. G.A. Tikhov, ed. Gosudarst. Nauch-Tekh. Gorno-Geol. Izd., Moscow, 1934.

Sharonov, V.V. Tables for the Calculation of Natural Light and Visibility. Izd. Akad. Nauk SSSR, 1945.

Shifrin, K.S., and Pyatkovskaya, I.N. On the indicatrix of luminance of natural surfaces. Trudy Glavnoi Geofiz. Observ. im. A. I. Voeikova, 68, 140, 1957.

Smirnov, N.V., and Dunin-Barkovskii, I.V. A Short Course of Mathematical Statistics for Technical Applications, Fizmatgiz, 1959.

Stepanov, B.I. Basic problems of the spectrometry of scattering media. Izvest. Akad. Nauk SSSR, Ser. Fiz., 21, 11, 1485, 1957.

Sytinskaya, N.N. A new optical characteristic of natural surfaces – the smoothness factor. Nauch. Byull. Leningrad. Gosudarst. Univ., 6, 6, 1946.

Tikhov, G.A. Izvest. Astronom. Obshchestva, October, 1911; November, 1914.

Tikhov, G.A. Color properties of foliage. Trudy Gosudarst. Nauch.-Issledovatel. Inst. Geodezii i Kartografii, Vol. 4, Leningrad, 1931.

Tikhov, G.A. Fluorescence of vegetation in infrared light. Vestnik Akad. Nauk Kazakh. SSR, No. 11 (44), 1948.

Tikhov, G.A. Principles of Visual and Photographic Photometry. Alma-Ata, 1950.

Tudorovskii, A.I. Theory of Optical Instruements, Pt. 2, Akad. Nauk SSSR, Moscow-Leningrad, 1952.

Van der Waerden, B.L. Mathematische Statistik [Russian translation]. IL, Moscow, 1960.

Vistelius, A.B. The problem of studying association in mineralogy and petrography. Zapiskii Vsesoyuz. Mineral. Obshchestva, Pt. 85, No. 1, 58, 1956.

Vistelius, A.B. Spectral luminance of sandy-silt rocks of the Aptian, Albian, and Cenomanian of Transcaspia. Geology of Transcaspia. Izd. Akad. Nauk SSSR, No. 1, 1958, pp. 31-67.

Vistelius, A.B. and Korobkov, I.A. On a new discovery of the Konkian horizon on Krasnovodsk plateau. Doklady Akad. Nauk SSSR, XC, No. 3, 1953.

Vistelius, A.B. and Sarmanov, O.V. On the correlation between percentage values. J. Geol. 69, 2, 1961.

Vistelius, A.B., and Yaroslavskaya, N.N. The main features of the color characterization of sandy-silt deposits of the terrigenous Cretaceous of Transcaspia. Doklady Akad. Nauk SSSR, XCV, 2, 367, 1954.

Volosov, D.S., and Tsivkin, M.V. Theory and Calculation of Optical Systems. Iskusstvo, Moscow, 1960.

Voronkova, N.M., Meleshko, K.E., Semenchenko, I.V., Snytkin, A.V., and Shishkina, T.A. A study of the spectral luminance of natural formations. Geodeziya i Kartografiya, No. 11, 1960.

Waldron, J.R., and Tellex, P.A. Reflectance of magnesium oxide. J. Opt. Soc. Am., 1, 1955.

Zaidel', A.I., Prokof'ev, V.K., and Raiskii, S.M. Tables of Spectral Lines. Gosudarst. Izd. Tekh-Teor. Lit., Moscow-Leningrad, 1952.

Zhidkova, Z.V. Reflection spectra of colored scattering objects. Zhur. Eksp. i Teoret. Fiz. 27, 4(10), 458, 1954.

Zoller, C.F. Photometrische Untersuchungen. Basel, 1859.

APPENDICES

APPENDIX I. Height of Sun at Each Hour of Local Time (from V.V. Sharonov, 1945)

Height of sun for latitude 35°

Date	Year Ord / Leap	0	1	2	3	4	5	6	7	8	9	10	11	12	13	14	15	16	17	18	19	20	21	22	23
January 1 / 2							−26°	−14°	−2°	8°	17°	25°	30°	31°	30	25°	18°	9°	−1°	−12°	−24°				
11 / 12							−26	−14	−2	8	17	25	31	33	32	27	20	11	0	−10	−23				
21 / 22							−26	−14	−2	8	18	26	32	35	34	29	22	13	2	−8	−21				
31 / 32							−25	−13	−1	10	20	28	34	38	37	32	25	15	5	−7	−19				
February 10 / 11							−23	−11	1	12	23	31	37	40	40	34	27	17	6	−5	−17	−30°			
20 / 21							−21	−9	3	15	25	34	40	44	43	33	29	19	8	−4	−16	−28			
March 2						−31°	−19	−7	6	17	27	37	44	47	46	40	32	22	11	−2	−14	−25			
12						−28	−16	−4	8	20	31	41	48	51	50	44	34	24	12	0	−12	−24			
22						−26	−14	−2	11	23	34	44	51	55	53	46	36	26	14	2	−11	−22			
April 1					−26°	−22	−10	2	14	26	38	49	57	60	57	50	40	28	16	4	−9	−20	−27°		
11					−24	−19	−7	5	17	30	42	52	60	64	60	52	42	30	17	5	−7	−19	−25		
21					−22	−16	−5	7	19	32	43	55	64	67	63	55	43	32	19	7	−5	−16	−23		
May 1					−20	−13	−2	9	21	34	45	56	66	70	65	56	45	33	20	8	−4	−14	−20		
11					−18	−11	−1	10	23	35	47	58	68	73	67	57	46	33	21	10	−2	−12	−19		
21					−17	−10	0	12	24	36	49	60	70	75	69	59	48	35	22	11	0	−10	−17		
31					−17	−9	1	13	25	37	50	61	71	76	70	60	49	36	24	12	1	−9	−17		
June 10				−25°	−18	−8	2	13	25	37	50	62	72	77	71	61	50	37	25	13	2	−8	−17	−25°	
20				−25	−19	−8	2	13	25	37	50	62	72	78	72	62	50	37	25	13	2	−8	−18	−25	
30				−25	−21	−9	1	12	24	36	49	60	71	77	72	61	49	37	25	13	2	−8	−19	−25	
July 10				−26	−22	−10	0	11	23	35	48	59	70	76	72	61	49	37	25	13	2	−9	−20	−25	
20					−24	−12	0	10	22	34	47	58	69	75	71	61	49	37	25	12	1	−10	−23		
30					−26	−13	−1	9	21	33	46	57	68	73	70	60	48	36	24	11	−1	−11	−26		
August 9					−28	−15	−3	8	20	32	44	56	66	70	67	58	46	34	22	10	−2	−13	−28		
19						−16	−5	7	19	31	43	54	64	67	64	56	44	32	20	8	−4	−15			
29						−18	−7	6	17	29	42	52	61	64	61	53	42	30	17	6	−6	−18			
September 8						−20	−8	5	16	28	40	50	58	61	58	50	39	27	15	3	−9	−21			
18						−22	−10	3	14	26	38	47	54	57	54	46	35	24	12	0	−12	−24			
28						−23	−11	1	13	25	36	45	51	54	50	42	32	21	9	−3	−15	−27			
October 8						−25°	−13	−1	11	23	33	42	48	49	45	38	29	18	7	−6	−18	−30			
18						−27	−15	−3	9	20	30	38	44	45	41	35	26	15	4	−8	−20				
28						−29	−17	−4	7	18	27	35	40	41	38	32	23	13	2	−11	−21				
November 7						−30	−18	−6	5	16	25	32	37	38	36	29	20	10	0	−13	−23				
17							−20	−8	4	14	23	30	35	36	34	28	19	9	−1	−14	−24				
27							−21	−9	2	12	21	28	33	34	32	26	18	8	−2	−14	−24				
December 7							−23	−11	0	10	19	26	31	32	30	25	18	8	−2	−14	−25				
17							−24	−12	−1	9	17	25	30	31	29	24	17	8	−2	−14	−25				
27							−26	−13	−2	8	16	24	29	31	29	24	17	8	−2	−13	−24				

148

APPENDIX I (cont'd). Height of Sun at Each Hour of Local Time (from V.V. Sharonov, 1945)

Height of sun for latitude 40°

Date (Ord./Leap)	0	1	2	3	4	5	6	7	8	9	10	11	12	13	14	15	16	17	18	19	20	21	22	23
Jan 1 / 2						−26	−15	−5	5	14	20	25	26	25	21	15	6	−3	−14	−25				
Jan 11 / 12						−26	−15	−5	5	14	21	26	28	26	23	16	8	−2	−13	−24				
Jan 21 / 22						−26	−15	−4	6	15	22	28	30	29	25	19	10	0	−10	−22				
Jan 31 / 32						−25	−14	−3	8	17	24	30	33	32	29	21	12	2	−8	−22				
Feb 10 / 11						−23	−12	−1	10	20	28	33	36	35	30	24	15	5	−6	−19				
Feb 20 / 21						−21	−10	2	12	22	30	36	39	38	34	26	17	7	−4	−16				
Mar 2					−30	−18	−7	4	15	25	34	40	42	41	37	29	20	9	−2	−14	−25			
Mar 12					−27	−16	−4	7	18	28	37	44	46	45	40	31	22	11	0	−12	−23			
Mar 22					−24	−13	−2	10	21	31	40	47	50	48	43	34	24	13	2	−10	−21			
Apr 1					−20	−9	2	14	25	35	45	52	55	53	46	37	27	15	4	−7	−18	−28		
Apr 11				−26	−16	−6	6	17	28	39	49	56	58	56	49	39	29	17	6	−5	−16	−26		
Apr 21				−23	−14	−3	8	19	31	42	52	59	62	59	52	42	31	19	8	−3	−14	−23		
May 1				−20	−11	−1	10	22	33	44	54	62	66	61	53	43	32	20	9	−2	−12	−21		
May 11				−18	−9	0	12	23	35	46	56	64	68	64	54	44	33	22	10	−1	−10	−20		
May 21			−23	−16	−7	2	13	24	36	47	58	66	70	65	57	46	35	23	12	1	−8	−17	−24	
May 31			−22	−15	−5	4	14	26	37	48	59	68	71	67	58	47	36	25	13	3	−6	−15	−22	
Jun 10			−21	−14	−5	4	15	26	38	49	60	69	72	69	59	48	37	26	14	4	−5	−14	−21	
Jun 20			−21	−14	−5	4	15	26	37	49	60	69	73	69	60	49	38	26	15	5	−5	−14	−21	
Jun 30			−21	−14	−6	4	14	25	37	48	59	68	73	69	60	49	37	26	15	5	−5	−14	−21	
Jul 10			−22	−15	−7	3	13	24	36	47	58	67	72	68	59	49	37	26	15	4	−6	−14	−22	
Jul 20			−24	−17	−9	1	12	23	34	46	56	66	70	67	58	48	36	25	14	3	−7	−15	−23	
Jul 30			−26	−19	−10	0	11	22	33	44	55	64	68	65	57	47	35	24	13	2	−8	−17	−24	
Aug 9				−21	−12	−2	9	20	32	43	54	62	66	63	55	45	33	22	11	0	−10	−20		
Aug 19				−23	−14	−3	8	19	30	41	52	60	63	60	53	43	31	20	8	−2	−13	−23		
Aug 29				−25	−16	−5	6	17	28	40	49	57	59	57	49	40	29	17	6	−5	−15	−25		
Sep 8				−27	−18	−7	4	15	27	38	46	54	56	54	46	37	27	15	4	−7	−18	−27		
Sep 18					−20	−9	2	13	25	35	44	50	52	49	42	33	23	11	0	−11	−22			
Sep 28					−22	−11	0	12	23	33	41	47	48	45	39	30	19	8	−3	−15	−26			
Oct 8					−24	−13	−2	10	20	30	38	43	44	41	35	26	16	6	−6	−17	−29			
Oct 18					−26	−15	−4	7	18	27	34	39	40	37	32	23	13	3	−9	−20				
Oct 28					−28	−17	−6	5	15	24	31	35	36	33	30	20	10	0	−12	−23				
Nov 7					−30	−18	−8	3	13	21	28	32	33	30	29	17	7	−3	−14	−25				
Nov 17						−20	−9	1	11	20	26	30	31	28	27	16	7	−4	−15	−25				
Nov 27						−21	−11	0	9	18	24	28	29	26	26	15	6	−4	−15	−25				
Dec 7						−23	−13	−2	7	16	22	26	27	25	24	15	5	−5	−16	−26				
Dec 17						−24	−14	−4	6	15	21	25	26	24	22	14	4	−5	−16	−27				
Dec 27						−25	−15	−5	5	14	20	24	26	24	21	15	5	−4	−15	−26				

149

APPENDIX I (con'd). Height of Sun at Each Hour of Local Time (from V.V. Sharonov, 1945)

Height of sun for latitude 45°

Date	0	1	2	3	4	5	6	7	8	9	10	11	12	13	14	15	16	17	18	19	20	21	22	23
January 1 / 2						−27°	−17°	−7°	2°	10°	16°	20°	21°	20°	17°	11°	3°	−6°	−16°	−26°				
11 / 12						−27	−17	−7	2	10	17	21	23	22	18	12	5	−4	−14	−24				
21 / 22						−26	−16	−6	3	11	18	23	25	24	20	14	7	−2	−12	−22				
31 / 32						−26	−15	−5	5	13	20	26	28	27	23	18	10	0	−10	−20				
February 10 / 11						−23	−12	−2	8	16	23	28	31	30	26	20	12	3	−7	−18	−28°			
20 / 21						−21	−10	0	10	19	26	32	34	33	29	23	14	5	−5	−16	−26			
March 2						−18°	−7	3	13	23	30	35	38	37	32	26	17	8	−3	−13	−24			
12						−15	−4	6	16	26	33	39	41	40	36	28	20	10	−1	−11	−21			
22						−12	−2	9	19	29	37	42	45	44	39	31	22	12	2	−9	−19	−28°		
April 1				−27°	−18°	−9	2	13	23	33	41	47	50	48	42	34	25	15	4	−7	−17	−26		
11				−23	−14	−4	6	17	27	36	45	51	54	52	45	37	27	17	6	−4	−14	−23		
21				−20	−12	−2	8	19	30	40	48	54	57	54	48	40	30	19	8	−2	−11	−20		
May 1			−24°	−17	−9	1	11	22	32	42	51	57	60	57	50	42	32	21	10	0	−9	−18	−24°	
11			−21	−15	−6	3	13	24	34	44	53	60	62	60	52	44	33	22	12	2	−7	−16	−22	−24°
21		−23°	−19	−13	−5	4	14	25	35	46	55	62	64	61	54	45	34	24	14	4	−6	−14	−19	−24
31		−22	−18	−12	−4	6	15	26	36	47	56	64	65	63	56	46	36	25	15	5	−4	−12	−18	−22
June 10		−21	−17°	−11	−3	7	16	27	37	48	57	65	67	64	57	47	37	25	16	6	−3	−11	−17°	−21
20		−20	−16	−10	−2	7	16	27	37	48	58	65	68	65	58	48	38	25	17	7	−2	−10	−16	−20
30		−20	−16	−11	−3	6	15	26	36	47	57	65	67	65	57	48	38	24	16	7	−2	−10	−14	−20
July 10		−21	−18°	−13	−4	5	14	25	35	46	56	64	66	64	56	47	37	24	16	6	−3	−11	−17°	−20
20		−23	−20	−15	−6	3	13	23	34	44	54	62	65	62	55	46	36	24	14	4	−5	−14	−20	
30			−21	−17	−7	2	12	22	33	43	52	60	63	61	54	45	35	23	13	3	−6	−15	−22	
August 9			−24°	−18°	−9	0	10	20	31	41	50	58	61	58	52	43	33	22	11	0	−8	−16	−23	
19				−20	−11	−2	9	19	29	40	48	55	58	55	49	41	30	20	9	−1	−11	−19		
29				−22	−13	−4	7	17	27	38	46	52	54	52	46	38	27	17	7	−4	−13	−22		
September 8				−24	−15	−5	5	15	25	35	44	49	51	49	43	35	24	15	4	−6	−16	−25		
18					−18	−8	2	13	23	32	40	45	47	45	39	30	21	11	0	−10	−20			
28					−20	−10	0	11	21	30	37	42	43	40	35	27	17	7	−3	−14	−24			
October 8					−23	−12	−2	8	18	27	34	38	39	36	31	23	14	5	−6	−17	−27			
18					−26	−15	−4	6	15	24	30	34	35	33	28	20	12	2	−9	−20				
28					−28	−17	−6	3	13	21	27	30	31	29	24	16	9	−1	−12	−23				
November 7						−19	−8	1	10	18	24	28	28	26	22	14	7	−3	−14	−24				
17						−21	−10	−1	8	16	22	26	26	24	20	12	4	−5	−15	−25				
27						−23	−12	−3	6	14	20	23	24	22	18	11	3	−6	−15	−26				
December 7						−25	−14	−5	4	12	18	21	22	21	17	10	2	−7	−16	−27				
17						−26	−15	−6	3	11	17	20	21	20	16	10	2	−7	−17	−27				
27						−27	−17	−7	2	10	16	20	21	20	16	9	2	−7	−17	−27				

APPENDIX I (cont'd) Height of Sun at Each Hour of Local Time (from V. V. Sharonov, 1945)

Height of sun for latitude 50°

Date (Ord./Leap)	0	1	2	3	4	5	6	7	8	9	10	11	12	13	14	15	16	17	18	19	20	21	22	23
January 1/2						−28°	−18°	−9°	−1°	5°	12°	15°	16°	15°	12°	6°	1°	−8°	−17°	−26°				
11/12						−27	−18	−9	0	7	13	16	18	17	14	8	2	−7	−16	−25				
21/22						−26	−17	−8	1	8	14	18	20	19	16	11	4	−4	−13	−23				
31/32						−25	−16	−7	2	10	16	21	23	23	19	14	7	−2	−11	−20				
February 10/11						−23	−13	−5	5	13	19	24	26	26	22	16	9	0	−8	−18	−28°			
20/21						−20	−10	−2	8	16	23	27	29	28	25	19	12	3	−6	−16	−25			
March 2				−24°	−27°	−17	−8	2	11	19	26	30	32	31	28	22	15	6	−3	−13	−22	−25°		
12				−20	−23	−14	−4	5	14	23	29	34	36	35	31	25	17	8	−1	−11	−20			
22				−18	−20	−11	−1	8	18	26	33	37	39	37	32	28	19	11	2	−8	−17			
April 1				−17	−16	−7	3	12	22	30	38	42	44	42	37	32	23	14	4	−5	−13	−23°	−23°	
11				−16	−12	−3	6	16	26	34	42	47	48	47	42	34	26	16	6	−3	−12	−20		
21			−23°	−15	−9	0	9	19	28	37	45	50	51	50	45	37	28	19	9	0	−9	−17		
May 1			−20	−13	−6	3	12	22	31	40	48	52	54	52	47	39	30	21	11	2	−6	−14	−20	−22°
11		−21°	−18	−10	−4	5	14	24	33	42	50	54	57	54	49	41	32	22	13	3	−4	−12	−18	−18
21		−20	−17	−7	−2	6	15	25	35	43	52	56	59	56	51	43	34	24	14	5	−2	−10	−15	−17
31		−18	−16	−7	−1	7	16	26	36	45	54	58	60	58	53	44	35	26	16	7	−1	−8	−13	
June 10	−19°		−14	−8	0	8	17	27	37	46	55	60	62	60	54	44	36	27	17	8	0	−7	−10	−16
20	−17		−12	−10	1	9	18	27	37	46	55	60	63	61	55	45	37	28	18	9	1	−6	−10	−15
30	−17		−12	−11	0	8	17	27	36	46	54	60	63	61	55	45	37	28	18	9	1	−6	−11	−15
July 10	−17	−16	−13	−14	−1	7	16	26	35	45	53	60	62	60	55	45	37	27	18	8	0	−7	−12	−16
20	−19	−18	−15	−16	−3	6	14	24	34	43	51	58	60	58	53	44	36	26	16	7	−1	−8	−14	−18
30		−20	−17	−19	−4	4	13	23	32	41	50	56	58	57	51	43	34	25	15	6	−2	−10	−16	−20
August 9			−20	−21	−6	2	11	21	30	39	47	53	56	54	48	41	32	22	13	3	−5	−13	−19	
19			−22	−25	−9	0	9	19	28	37	45	51	53	51	46	38	29	20	10	1	−8	−16	−22	
29					−11	−2	7	16	26	35	42	47	49	47	42	35	26	16	7	−2	−11	−19		
September 8					−13	−4	5	14	24	33	40	45	46	45	40	32	24	14	5	−5	−14	−21		
18					−17	−7	2	12	21	29	36	41	42	40	35	27	20	10	1	−9	−18	−26		
28					−19	−10	0	10	18	26	33	37	38	36	31	24	16	6	−3	−13	−21			
October 8					−22	−12	−2	7	16	23	30	33	34	32	27	20	13	4	−6	−16	−25			
18					−24	−15	−5	4	13	20	26	30	30	28	24	17	9	1	−9	−19	−28			
28					−27	−17	−8	2	10	17	23	26	26	24	20	14	6	−3	−12	−22				
November 7						−20	−10	−1	7	14	20	23	23	22	17	11	3	−6	−15	−25				
17						−21	−12	−3	5	12	18	21	22	20	15	9	2	−8	−17	−27				
27						−24	−14	−5	3	10	15	18	19	18	13	7	0	−9	−18	−27				
December 7						−25	−16	−7	1	8	14	16	17	16	12	6	0	−9	−18	−27				
17						−26	−17	−8	0	6	13	15	16	15	11	6	−1	−9	−18	−28				
27						−27	−18	−9	−1	5	12	15	16	15	12	6	0	−8	−17	−27				

APPENDIX I (cont'd) Height of Sun at Each Hour of Local Time (from V.V. Sharonov, 1945)

Height of sun for latitude 55°

Date	Ord.	Leap	0	1	2	3	4	5	6	7	8	9	10	11	12	13	14	15	16	17	18	19	20	21	22	23
January	1	2							−19	−10	−4	3	8	11	11	11	8	3	−3	−11	−18					
	11	12							−19	−10	−3	4	9	12	13	13	10	5	−1	−9	−16	−21				
	21	22							−18	−10	−2	5	10	14	15	15	12	7	1	−6	−14	−23				
	31	32						−20	−16	−8	0	7	12	16	18	17	15	10	4	−3	−12	−20				
February	10	11						−22	−13	−5	3	10	15	19	21	20	17	13	6	−1	−10	−18	−26			
	20	21						−19	−11	−2	5	13	18	22	24	23	20	16	9	1	−7	−16	−24			
March	2							−16	−8	1	8	16	21	26	27	27	24	19	12	4	−4	−13	−21			
	12						−25	−12	−4	4	12	20	25	29	31	31	27	22	15	7	−2	−10	−18			
	22					−24	−21	−9	−1	7	16	24	29	33	35	34	30	25	17	10	2	−7	−15	−26		
April	1					−20	−17	−5	3	11	20	28	34	38	40	39	34	29	21	13	4	−4	−12	−22		
	11				−22	−16	−13	−2	7	15	24	32	38	42	44	42	38	32	24	15	7	−2	−10	−20	−22	
	21			−19	−19	−13	−10	1	10	18	27	34	41	45	47	45	41	34	27	18	10	1	−7	−16	−19	−22
May	1		−20	−16	−15	−10	−7	4	13	21	30	38	44	48	50	48	43	37	29	20	12	4	−4	−13	−16	−19
	11		−18	−13	−13	−8	−3	6	15	23	32	40	46	51	53	50	45	39	31	22	13	6	−2	−11	−13	−16
	21		−14	−11	−10	−5	−1	8	17	25	34	42	48	54	55	53	48	41	33	24	15	8	0	−6	−10	−14
	31		−12	−10	−8	−3	3	10	19	27	35	43	50	56	57	55	50	42	35	26	17	10	2	−4	−8	−12
June	10		−11	−10	−7	−2	4	11	19	27	36	44	51	56	57	56	51	43	36	27	18	11	3	−3	−7	−11
	20		−11	−10	−7	−2	4	11	19	28	36	44	51	56	58	56	52	44	36	28	19	11	4	−2	−7	−10
	30		−12	−12	−7	−3	3	10	18	27	35	44	51	56	58	56	52	44	36	28	19	11	4	−2	−7	−10
July	10		−12	−14	−9	−4	2	9	17	25	33	43	50	55	57	55	51	43	35	27	18	10	3	−3	−8	−11
	20		−14	−15	−11	−6	0	7	15	24	32	41	48	53	55	54	49	42	34	26	17	9	2	−5	−10	−13
	30		−16	−18	−12	−8	−1	6	14	22	31	39	46	51	53	52	47	40	33	24	15	7	0	−6	−12	−14
August	9		−19	−21	−15	−10	−4	4	12	21	29	37	44	49	51	49	45	38	30	22	13	5	−3	−9	−15	−17
	19				−18	−13	−6	2	10	19	27	35	42	46	48	47	42	36	28	20	11	3	−5	−12	−17	−21
	29				−20	−15	−8	−1	7	16	24	32	38	42	44	43	38	32	24	16	7	−1	−8	−15	−20	
September	8				−23	−17	−11	−5	5	14	22	30	36	40	41	40	35	29	21	13	4	−4	−12	−19	−24	
	18					−21	−14	−6	2	11	19	26	32	36	36	35	31	25	17	9	0	−8	−16	−23		
	28					−24	−17	−9	0	8	16	23	28	32	33	31	27	21	13	5	−3	−12	−20			
October	8						−20	−12	−3	6	13	20	25	28	29	27	23	18	10	2	−6	−15	−23			
	18						−23	−14	−6	2	10	16	22	25	26	24	20	14	7	−1	−10	−18	−26			
	28						−26	−17	−9	0	7	13	18	21	22	20	16	10	3	−4	−13	−22				
November	7							−20	−11	−3	4	10	15	18	18	16	13	7	0	−7	−16	−24				
	17							−22	−13	−5	2	8	13	16	16	14	11	5	−1	−9	−17	−26				
	27							−24	−15	−7	0	6	11	13	14	12	9	3	−3	−10	−18					
December	7							−26	−17	−9	−2	4	8	11	12	11	8	2	−4	−11	−19					
	17							−27	−18	−10	−3	3	7	10	11	10	7	2	−4	−12	−20					
	27								−19	−11	−4	2	8	11	11	10	8	3	−3	−11	−19					

APPENDIX I. (cont'd) Height of Sun at Each Hour of Local Time (from V. V. Sharonov, 1945)

Height of sun for latitude 60°

Date	Ord.	Leap	0	1	2	3	4	5	6	7	8	9	10	11	12	13	14	15	16	17	18	19	20	21	22	23
January	1	2							−20°	−13°	−7°	−1°	3°	6°	6°	6°	3°	−4°	−6°	−12°	−20°					
	11	12							−20	−13	−6	0	4	7	8	8	5	1	−5	−11	−18					
	21	22							−19	−12	−5	1	6	9	10	10	7	3	−2	−8	−15	−23°				
	31	32						−24°	−17	−10	−3	3	8	12	13	13	10	6	1	−6	−13	−20				
February	10	11						−16	−14	−7	0	6	11	15	16	16	13	9	3	−3	−10	−18	−25°			
	20	21						−12	−11	−4	3	9	14	18	19	19	16	12	6	0	−8	−15	−22			
March	2						−22°	−16	−8	−1	6	13	18	21	22	22	19	15	10	3	−5	−12	−20			
	12					−25°	−19	−12	−4	3	10	16	21	24	26	26	23	18	13	6	−2	−9	−16	−22°		
	22					−21	−15	−8	−1	6	14	20	25	28	30	29	26	21	15	8	1	−6	−13	−19	−22°	
April	1				−22°	−17	−11	−4	3	11	19	25	30	33	35	34	31	26	19	12	5	−3	−10	−16	−20	
	11			−21°	−18	−13	−7	0	8	15	22	29	34	37	38	38	34	29	22	15	8	0	−7	−13	−17	−21°
	21		−19°	−18	−14	−10	−4	3	10	18	25	32	37	40	42	41	37	32	25	18	10	3	−4	−10	−14	−18
May	1		−15	−14	−11	−6	0	6	13	21	28	35	40	43	45	44	40	34	28	20	13	5	−3	−7	−11	−14
	11		−13	−11	−9	−4	2	8	15	23	30	37	42	46	48	46	42	36	30	22	15	7	1	−5	−9	−12
	21		−10	−8	−6	−1	4	10	17	25	32	39	44	48	50	48	44	38	32	24	17	10	3	−3	−6	−9
	31		−8	−6	−4	2	6	12	19	27	34	41	46	50	52	50	46	40	34	26	19	12	5	0	−4	−7
June	10		−7	−6	−3	1	6	13	20	27	35	42	47	51	53	51	47	41	35	27	20	12	6	1	−3	−6
	20		−6	−5	−3	1	7	13	20	28	35	42	48	52	53	52	48	42	36	28	21	13	7	1	−2	−5
	30		−7	−6	−4	1	6	12	20	27	34	41	47	51	53	51	48	42	35	28	20	13	7	1	−3	−5
July	10		−8	−7	−5	0	5	11	19	26	33	40	47	50	52	50	47	41	34	27	19	12	6	0	−4	−7
	20		−10	−9	−6	−2	3	10	17	24	31	38	45	48	50	49	45	40	33	26	18	11	4	−1	−5	−8
	30		−12	−10	−8	−4	1	8	15	22	30	37	43	46	49	48	44	38	31	24	17	9	2	−3	−7	−10
August	10		−14	−13	−11	−7	−1	6	13	20	28	34	40	44	46	44	41	35	29	21	14	7	0	−6	−10	−13
	20		−17	−16	−13	−9	−3	3	11	18	26	32	37	41	43	42	38	33	26	19	12	4	−2	−9	−13	−16
	30			−20	−16	−12	−6	1	8	15	23	29	34	38	39	38	34	29	23	15	8	1	−6	−12	−16	−20
September	9				−19	−14	−8	−2	5	13	20	26	31	35	35	35	31	26	20	13	5	−2	−9	−15	−20	
	19				−23	−18	−12	−5	2	10	16	22	28	31	32	30	27	22	16	9	1	−6	−13	−20		
	29					−22	−15	−8	0	7	14	20	24	27	28	27	23	18	12	6	−3	−10	−17	−24		
October	8					−25	−18	−11	−4	4	11	16	21	24	24	23	19	14	8	2	−6	−14	−21			
	18						−21	−14	−7	1	7	13	17	20	20	19	16	11	5	−2	−10	−18	−24			
	28						−25	−17	−9	−2	4	10	14	15	16	15	12	7	1	−6	−14	−21				
November	7							−20	−13	−6	0	6	10	12	13	11	8	3	−3	−9	−17	−24				
	17							−22	−15	−8	−2	4	8	10	11	9	6	2	−5	−11	−18	−25				
	27							−24	−17	−10	−4	2	6	8	8	7	4	0	−6	−12	−20					
December	7							−26	−18	−11	−5	0	4	6	7	6	3	−1	−7	−13	−20					
	17								−20	−12	−6	−1	3	5	6	5	2	−2	−7	−14	−21					
	27								−20	−13	−7	−1	3	6	6	6	2	−1	−6	−13	−20					

APPENDIX I (cont'd). Height of Sun for Each Hour of Local Time (from V.V. Sharonov, 1945)

Height of sun for latitude 65°

Date (Ord./Leap)	0	1	2	3	4	5	6	7	8	9	10	11	12	13	14	15	16	17	18	19	20	21	22	23
January 1 / 2							−21°	−15°	−10°	−5°	−1°	1°	2°	1°	−1°	−4°	−9°	−14°	−20°					
January 11 / 12							−20	−14	−9	−4	0	2	3	2	0	−3	−8	−13	−18	−26				
January 21 / 22						−24°	−19	−13	−8	−2	1	4	5	4	3	−1	−5	−11	−16	−23				
January 31 / 32						−21	−17	−11	−5	0	4	7	8	8	6	2	−2	−8	−14	−20				
February 10 / 11						−18	−14	−8	−2	3	7	10	11	11	8	5	0	−5	−11	−18	−24°			
February 20 / 21						−14	−11	−5	1	6	10	13	14	14	11	8	3	−2	−8	−15	−21			
March 2					−21°	−10	−8	−2	4	9	13	15	17	17	15	11	6	1	−5	−12	−18	−23°		
March 12				−21°	−17	−7	−4	2	8	13	17	20	21	21	19	15	10	4	−2	−8	−14	−20		
March 22			−22°	−18	−14	−3	−1	6	13	17	21	24	25	24	22	18	13	7	1	−5	−11	−16	−20°	
April 1		−20°	−18	−14	−9	1	4	10	17	22	26	29	30	29	27	23	17	11	5	−2	−8	−13	−17	−20°
April 11	−18°	−16	−13	−10	−4	4	8	13	20	25	30	33	34	33	30	26	20	14	8	1	−4	−10	−13	−16
April 21	−14	−13	−10	−6	−2	8	11	17	23	29	33	36	37	36	33	29	23	17	11	4	−1	−6	−10	−13
May 1	−10	−9	−6	−3	2	10	14	20	26	31	36	39	40	39	36	32	26	20	13	7	2	−3	−7	−10
May 11	−8	−7	−4	0	4	12	16	22	29	34	38	42	43	41	38	34	28	22	15	9	4	−1	−4	−7
May 21	−5	−4	−2	2	6	14	18	24	32	36	41	44	45	43	40	36	31	24	18	11	6	1	−2	−4
May 31	−3	−2	0	4	8	15	20	26	34	38	43	46	47	45	42	38	34	26	20	13	8	4	0	−2
June 10	−2	−1	1	4	9	15	21	27	34	39	44	47	48	46	43	39	34	27	21	14	9	5	0	−1
June 20	−2	−1	2	5	10	14	21	27	35	39	45	47	48	46	44	39	34	28	21	15	9	5	1	0
June 30	−2	−1	1	5	9	13	21	27	34	39	44	47	48	47	44	39	34	28	21	15	10	5	1	−1
July 10	−3	−2	0	4	8	12	20	26	32	38	43	46	47	45	43	38	33	26	20	14	9	4	0	−2
July 20	−4	−4	−2	2	6	10	18	24	30	36	41	44	45	44	42	37	32	25	19	13	7	3	−1	−4
July 30	−6	−6	−3	0	4	8	16	22	28	34	39	42	43	43	40	35	30	23	17	11	5	1	−3	−5
August 9	−9	−9	−6	−3	2	5	14	20	26	32	36	39	41	41	37	33	27	21	15	8	3	−2	−6	−9
August 19	−12	−11	−9	−6	0	2	11	18	24	29	34	36	38	37	34	30	24	18	12	6	0	−5	−9	−11
August 29	−15	−14	−12	−9	−3	0	8	15	21	26	30	33	34	33	30	26	21	15	8	2	−3	−9	−12	−14
September 8	−18	−17	−15	−11	−6	−4	6	12	18	23	27	30	31	30	27	23	18	11	5	−1	−6	−12	−15	−17
September 18			−20	−15	−10	−7	2	8	14	19	23	26	27	26	23	19	14	7	1	−5	−10	−16	−17	−20
September 28				−19	−13	−10	−1	5	11	16	20	22	23	23	19	15	9	3	−3	−9	−15	−20		
October 8				−22	−16	−15	−4	2	7	13	16	18	19	18	15	11	6	0	−7	−13	−19			
October 18					−20	−17	−7	−1	4	9	13	15	15	14	11	7	2	−4	−10	−16	−22			
October 28					−23	−20	−11	−4	1	6	9	11	11	10	7	3	−2	−7	−14	−20				
November 7						−22	−13	−7	−2	2	6	8	8	7	4	0	−5	−11	−17	−23				
November 17						−24	−15	−9	−4	0	3	6	6	5	2	−1	−6	−12	−18	−24				
November 27							−18	−12	−6	−2	1	3	4	3	0	−3	−8	−13	−19					
December 7							−20	−14	−8	−3	0	1	2	1	−1	−5	−9	−14	−20					
December 17							−21	−15	−10	−4	−2	0	2	0	−2	−6	−10	−15	−21					
December 27							−21	−15	−10	−5	−2	1	2	1	−2	−5	−10	−14	−20					

APPENDIX I (cont'd). Height of Sun for Each Hour of Local Time (from V. V. Sharonov, 1945)

Height of sun for latitude 70°

Date	Year Ord./Leap	0	1	2	3	4	5	6	7	8	9	10	11	12	13	14	15	16	17	18	19	20	21	22	23
January	1 / 2							−22°	−17°	−13°	−9°	−6°	−4	−3	−4	−7°	−8°	−12°	−16°	−21°					
	11 / 12							−21	−16	−12	−8	−5	−2	−1	−2	−4	−7	−12	−15	−20					
	21 / 22							−20	−15	−10	−6	−3	0	0	0	−2	−4	−10	−12	−17					
	31 / 32							−18	−12	−8	−4	−1	2	3	3	1	−2	−8	−10	−15					
February	10 / 11						−22°	−15	−9	−5	−1	3	5	6	6	4	1	−5	−7	−12	−22				
	20 / 21					−21°	−20	−11	−6	−2	3	6	8	9	9	7	4	−2	−4	−9	−20	−21°			
March	2				−22°	−18	−17	−8	−3	2	6	10	12	12	12	11	8	2	−1	−6	−17	−18	−22°		
	12			−21°	−18	−14	−13	−4	1	5	10	14	15	16	16	14	11	5	3	−2	−13	−14	−18	−21°	
	22	−20°	−19°	−17	−14	−10	−9	0	5	9	14	17	19	20	20	18	15	9	6	1	−9	−10	−14	−17	−19°
April	1	−16	−15	−13	−10	−6	−5	4	9	14	18	22	24	25	24	22	19	14	10	5	−5	−6	−10	−13	−15
	11	−12	−11	−9	−6	−2	−1	8	13	18	22	26	28	29	28	26	22	18	13	8	−1	−2	−6	−9	−11
	21	−9	−8	−6	−3	1	3	11	16	21	26	29	30	32	31	29	26	21	16	11	3	1	−3	−6	−8
May	1	−5	−4	−1	0	4	6	13	18	23	29	31	33	35	34	31	29	23	19	14	6	4	0	−1	−4
	11	−2	−2	0	3	7	9	15	20	25	31	34	36	37	36	33	30	25	21	16	9	7	3	0	−2
	21	0	1	3	6	9	11	17	22	27	32	37	38	40	39	36	33	27	23	18	11	9	6	3	1
	31	2	3	5	8	11	14	19	24	29	33	39	40	41	41	38	35	29	25	20	14	11	8	5	3
June	10	3	3	6	9	11	16	20	25	31	34	39	41	42	42	39	36	31	26	21	16	11	9	6	3
	20	3	4	6	9	12	16	21	26	32	36	40	42	43	43	40	37	32	28	22	16	12	9	6	4
	30	3	4	6	8	12	17	21	26	32	36	40	42	43	43	40	37	32	28	22	17	12	8	6	4
July	10	2	3	5	7	11	17	20	25	31	35	39	41	42	42	39	36	31	27	21	17	11	7	5	3
	20	1	1	3	6	9	16	19	23	29	33	37	39	40	40	37	34	29	25	20	16	9	6	3	1
	30	−1	0	1	4	7	14	17	22	27	31	35	38	39	38	35	32	27	23	18	14	7	4	1	0
August	9	−4	−3	−2	1	5	12	14	19	24	28	32	35	36	35	33	29	24	20	15	12	5	1	−2	−3
	19	−7	−6	−5	−2	2	9	12	17	21	24	30	32	33	32	30	26	21	17	12	9	2	−2	−5	−6
	29	−10	−9	−8	−5	−1	6	9	14	19	22	26	28	29	28	26	22	19	14	9	6	−1	−5	−8	−9
September	8	−13	−12	−11	−7	−3	4	6	11	16	20	23	25	26	25	23	20	16	11	5	4	−3	−7	−11	−12
	18	−17	−17	−14	−12	−7	1	2	7	12	16	19	21	22	21	19	15	12	6	1	1	−7	−12	−14	−17
	28		−21	−18	−16	−11	−3	−1	4	8	12	16	17	18	17	15	11	8	2	−4	−3	−11	−16	−18	
October	8			−22°	−18	−14	−6	−5	0	5	9	12	13	14	13	11	7	5	−1	−7	−6	−14	−18	−22°	
	18				−22°	−18	−10	−8	−3	1	5	8	9	11	9	7	4	1	−5	−10	−10	−18	−22°		
	28					−22°	−13	−12	−6	−2	2	4	6	6	5	3	0	−2	−9	−14	−13	−22°			
November	7						−16	−14	−10	−5	−2	1	3	2	2	0	−3	−5	−12	−17	−16				
	17						−20	−16	−12	−7	−4	−1	1	1	0	−2	−5	−7	−13	−18	−20				
	27						−21	−19	−14	−10	−6	−4	−1	−1	−2	−4	−7	−10	−15	−19	−21				
December	7						−24	−20	−16	−11	−8	−5	−3	−2	−3	−5	−8	−11	−17	−20	−24				
	17							−21	−17	−12	−9	−6	−4	−3	−4	−6	−9	−12	−18	−22					
	27							−22	−17	−12	−9	−6	−4	−3	−4	−7	−9	−12	−17	−22					

APPENDIX II. Chebyshev Numbers for Odd Numbers of Divisions (from Mitropol'skii, 1961)

No. of divisions	φ_1	$3\varphi_2$	$\frac{5}{6}\varphi_3$
1	-4	28	-14
2	-3	7	7
3	-2	-8	13
4	-1	-17	9
5	0	-20	0
6	1	-17	-9
7	2	-8	-13
8	3	7	-7
9	4	28	14
9	60	2772	990
K_0	K_1	K_2	K_3

No. of divisions	φ_1	φ_2	$\frac{5}{6}\varphi_3$
1	-5	15	-30
2	-4	6	6
3	-3	-1	22
4	-2	-6	23
5	-1	-9	14
6	0	-10	0
7	1	-9	-14
8	2	-6	-23
9	3	-1	-22
10	4	6	-6
11	5	15	30
11	110	858	4290
K_0	K_1	K_2	K_3

No. of divisions	φ_1	φ_2	$\frac{1}{6}\varphi_3$
1	-6	22	-11
2	-5	11	0
3	-4	2	6
4	-3	-5	8
5	-2	-10	7
6	-1	-13	4
7	0	-14	0
8	1	-13	-4
9	2	-10	-7
10	3	-5	-8
11	4	2	-6
12	5	11	0
13	6	22	11
13	182	2002	572
K_0	K_1	K_2	K_3

No. of divisions	φ_1	$3\varphi_2$	$\frac{5}{6}\varphi_3$
1	-7	91	-91
2	-6	52	-13
3	-5	19	35
4	-4	-8	58
5	-3	-29	61
6	-2	-44	49
7	-1	-53	27
8	0	-56	0
9	1	-53	-27
10	2	-44	-49
11	3	-29	-61
12	4	-8	-58
13	5	19	-35
14	6	52	13
15	7	91	91
15	280	37,128	39,780
K_0	K_1	K_2	K_3

No. of divisions	φ_1	φ_2	$\frac{1}{6}\varphi_3$
1	-8	40	-28
2	-7	25	-7
3	-6	12	7
4	-5	1	15
5	-4	-8	18
6	-3	-15	17
7	-2	-20	13
8	-1	-23	7
9	0	-24	0
10	1	-23	-7
11	2	-20	-13
12	3	-15	-17
13	4	-8	-18
14	5	1	-15
15	6	12	-7
16	7	25	7
17	8	40	28
17	408	7752	3876
K_0	K_1	K_2	K_3

No. of divisions	φ_1	φ_2	$\frac{5}{6}\varphi_3$
1	-9	51	-204
2	-8	34	-68
3	-7	19	28
4	-6	6	89
5	-5	-5	120
6	-4	-14	126
7	-3	-21	112
8	-2	-26	83
9	-1	-29	44
10	0	-30	0
11	1	-29	-44
12	2	-26	-83
13	3	-21	-112
14	4	-14	-126
15	5	-5	-120
16	6	6	-89
17	7	19	-28
18	8	34	68
19	9	51	204
19	570	13,566	213,180
K_0	K_1	K_2	K_3

No. of divisions	φ_1	$3\varphi_2$	$\frac{5}{6}\varphi_3$
1	-10	190	-285
2	-9	133	-114
3	-8	82	12
4	-7	37	98
5	-6	-2	149
6	-5	-35	170
7	-4	-62	166
8	-3	-83	142
9	-2	-98	103
10	-1	-107	54
11	0	-110	0
12	1	-107	-54
13	2	-98	-103
14	3	-83	-142
15	4	-62	-166
16	5	-35	-170
17	6	-2	-149
18	7	37	-98
19	8	82	-12
20	9	133	114
21	10	190	285
21	770	201 894	432 630
K_0	K_1	K_2	K_3

No. of divisions	φ_1	φ_2	$\frac{1}{6}\varphi_3$
1	-11	77	-77
2	-10	56	-35
3	-9	37	-3
4	-8	20	20
5	-7	5	35
6	-6	-8	43
7	-5	-19	45
8	-4	-28	42
9	-3	-35	35
10	-2	-40	25
11	-1	-43	13
12	0	-44	0
13	1	-43	-13
14	2	-40	-25
15	3	-35	-35
16	4	-28	-42
17	5	-19	-45
18	6	-8	-43
19	7	5	-35
20	8	20	-20
21	9	37	3
22	10	56	35
23	11	77	77
23	1012	35 420	32 890
K_0	K_1	K_2	K_3

APPENDIX II (cont'd) Chebyshev Numbers for Odd Numbers of Divisions (from Mitropol'skii, 1961)

No. of divisions	Φ_1	Φ_2	$\frac{5}{6}\Phi_3$
1	−12	92	−506
2	−11	69	−253
3	−10	48	−55
4	−9	29	93
5	−8	12	196
6	−7	−3	259
7	−6	−16	287
8	−5	−27	285
9	−4	−36	258
10	−3	−43	211
11	−2	−48	149
12	−1	−51	77
13	0	−52	0
14	1	−51	−77
15	2	−48	−149
16	3	−43	−211
17	4	−36	−258
18	5	−27	−285
19	6	−16	−287
20	7	−3	−259
21	8	12	−196
22	9	29	−93
23	10	48	55
24	11	69	253
25	12	92	506
25	1300	53 820	1 480 050
K_0	K_1	K_2	K_3

No. of divisions	Φ_1	$3\Phi_2$	$\frac{1}{6}\Phi_3$
1	−13	325	−130
2	−12	250	−70
3	−11	181	−22
4	−10	118	15
5	−9	61	42
6	−8	10	60
7	−7	−35	70
8	−6	−74	73
9	−5	−107	70
10	−4	−134	62
11	−3	−155	50
12	−2	−170	35
13	−1	−179	18
14	0	−182	0
15	1	−179	−18
16	2	−170	−35
17	3	−155	−50
18	4	−134	−62
19	5	−107	−70
20	6	−74	−73
21	7	−35	−70
22	8	10	−60
23	9	61	−42
24	10	118	−15
25	11	181	22
26	12	250	70
27	13	325	130
27	1638	712 530	101 790
K_0	K_1	K_2	K_3

No. of divisions	Φ_1	Φ_2	$\frac{5}{6}\Phi_3$
1	−14	126	−819
2	−13	99	−468
3	−12	74	−182
4	−11	51	44
5	−10	30	215
6	−9	11	336
7	−8	−6	412
8	−7	−21	448
9	−6	−34	449
10	−5	−45	420
11	−4	−54	366
12	−3	−61	292
13	−2	−66	203
14	−1	−69	104
15	0	−70	0
16	1	−69	−104
17	2	−66	−203
18	3	−61	−292
19	4	−54	−366
20	5	−45	−420
21	6	−34	−419
22	7	−21	−448
23	8	−6	−412
24	9	11	−336
25	10	30	−215
26	11	51	−44
27	12	74	182
28	13	99	468
29	14	126	819
29	2030	113 274	4 207 320
K_0	K_1	K_2	K_3

No. of divisions	Φ_1	Φ_2	$\frac{5}{6}\Phi_3$
1	−15	145	−1015
2	−14	116	−609
3	−13	89	−273
4	−12	64	−2
5	−11	41	209
6	−10	20	365
7	−9	1	471
8	−8	−16	532
9	−7	−31	553
10	−6	−44	539
11	−5	−55	495
12	−4	−64	426
13	−3	−71	337
14	−2	−76	233
15	−1	−79	119
16	0	−80	0
17	1	−79	−119
18	2	−76	−233
19	3	−71	−337
20	4	−64	−426
21	5	−55	−495
22	6	−44	−539
23	7	−31	−553
24	8	−16	−532
25	9	1	−471
26	10	20	−365
27	11	41	−209
28	12	64	2
29	13	89	273
30	14	116	609
31	15	145	1015
31	2480	158 224	6 724 520
K_0	K_1	K_2	K_3

APPENDIX III. Significance a of Deviation of χ^2 from Zero Up to a Given Value for f degrees of Freedom

f	0.995	0.990	0.975	0.950	0.900	0.750	0.500	0.250	0.100	0.050	0.025	0.010	0.005
1	0.0^4393	0.0^3157	0.0^3982	0.0^2393	0.0158	0.102	0.455	1.32	2.71	3.84	5.02	6.63	7.88
2	0.0100	0.0201	0.0506	0.103	0.211	0.575	1.39	2.77	4.61	5.99	7.38	9.21	10.6
3	0.0717	0.115	0.216	0.352	0.584	1.21	2.37	4.11	6.25	7.81	9.35	11.3	12.8
4	0.207	0.297	0.484	0.711	1.06	1.92	3.36	5.39	7.78	9.49	11.1	13.3	14.9
5	0.412	0.554	0.831	1.15	1.61	2.67	4.35	6.63	9.24	11.1	12.8	15.1	16.7
6	0.676	0.872	1.24	1.64	2.20	3.45	5.35	7.84	10.6	12.6	14.4	16.8	18.5
7	0.989	1.24	1.69	2.17	2.83	4.25	6.35	9.04	12.0	14.1	16.0	18.5	20.3
8	1.34	1.65	2.18	2.73	3.49	5.07	7.34	10.2	13.4	15.5	17.5	20.1	22.0
9	1.73	2.09	2.70	3.33	4.17	5.90	8.34	11.4	14.7	16.9	19.0	21.7	23.6
10	2.16	2.56	3.25	3.94	4.87	6.74	9.34	12.5	16.0	18.3	20.5	23.2	25.2
11	2.60	3.05	3.82	4.57	5.58	7.58	10.3	13.7	17.3	19.7	21.9	24.7	26.8
12	3.07	3.57	4.40	5.23	6.30	8.44	11.3	14.8	18.5	21.0	23.3	26.2	28.3
13	3.57	4.11	5.01	5.89	7.04	9.30	12.3	16.0	19.8	22.4	24.7	27.7	29.8
14	4.07	4.66	5.63	6.57	7.79	10.2	13.3	17.1	21.1	23.7	26.1	29.1	31.3
15	4.60	5.23	6.26	7.26	8.55	11.0	14.3	18.2	22.3	25.0	27.5	30.6	32.8
16	5.14	5.81	6.91	7.96	9.31	11.9	15.3	19.4	23.5	26.3	28.8	32.0	34.3
17	5.70	6.41	7.56	8.67	10.1	12.8	16.3	20.5	24.8	27.6	30.2	33.4	35.7
18	6.26	7.01	8.23	9.39	10.9	13.7	17.3	21.6	26.0	28.9	31.5	34.8	37.2
19	6.84	7.63	8.91	10.1	11.7	14.6	18.3	22.7	27.2	30.1	32.9	36.2	38.6
20	7.43	8.26	9.59	10.9	12.4	15.5	19.3	23.8	28.4	31.4	34.2	37.6	40.0
21	8.03	8.90	10.3	11.6	13.2	16.3	20.3	24.9	29.6	32.7	35.5	38.9	41.4
22	8.64	9.54	11.0	12.3	14.0	17.2	21.3	26.0	30.8	33.9	36.8	40.3	42.8
23	9.26	10.2	11.7	13.1	14.8	18.1	22.3	27.1	32.0	35.2	38.1	41.6	44.2
24	9.89	10.9	12.4	13.8	15.7	19.0	23.3	28.2	33.2	36.4	39.4	43.0	46.6
25	10.5	11.5	13.1	14.6	16.5	19.9	24.3	29.3	34.4	37.7	40.6	44.3	46.9
26	11.2	12.2	13.8	15.4	17.3	20.8	25.3	30.4	35.6	38.9	41.9	45.6	48.3
27	11.8	12.9	14.6	16.2	18.1	21.7	26.3	31.5	36.7	40.1	43.2	47.0	49.6
28	12.5	13.6	15.3	16.9	18.9	22.7	27.3	32.6	37.9	41.3	44.5	48.3	51.0
29	13.1	14.3	16.0	17.7	19.8	23.6	28.3	33.7	39.1	42.6	45.7	49.6	52.3
30	13.8	15.0	16.8	18.5	20.6	24.5	29.3	34.8	40.3	43.8	47.0	50.9	53.7
40	20.7	22.2	24.4	26.5	29.1	33.7	39.3	46.6	51.8	55.8	59.3	63.7	66.8
50	28.0	29.7	32.4	34.8	37.7	42.9	49.3	56.3	63.2	67.5	71.4	76.2	79.5
60	35.5	37.5	40.5	43.2	46.5	52.3	59.3	67.0	74.4	79.1	83.3	88.4	92.0

Note. The index figure above the zero indicates the number of zeros between the decimal point and the first significant figure. For instance, $0.0^41 = 0.00001$.

APPENDIX IV. Table of Values of $Z = 1.15129 \log \dfrac{1 + r}{1 - r}$

r	0	1	2	3	4	5	6	7	8	9
0.0	0.0000	0.0100	0.0200	0.0300	0.0400	0.0500	0.0601	0.0701	0.0802	0.0902
0.1	0.1003	0.1104	0.1206	0.1307	0.1409	0.1511	0.1614	0.1717	0.1820	0.1923
0.2	0.2027	0.2132	0.2237	0.2342	0.2448	0.2554	0.2661	0.2769	0.2877	0.2986
0.3	0.3095	0.3205	0.3316	0.3428	0.3541	0.3654	0.3769	0.3884	0.4001	0.4118
0.4	0.4236	0.4356	0.4477	0.4599	0.4722	0.4847	0.4973	0.5101	0.5230	0.5361
0.5	0.5493	0.5627	0.5763	0.5901	0.6042	0.6184	0.6328	0.6475	0.6625	0.6777
0.6	0.6931	0.7089	0.7250	0.7414	0.7582	0.7753	0.7928	0.8107	0.8291	0.8480
0.7	0.8673	0.8872	0.9076	0.9287	0.9505	0.9730	0.9962	1.0203	1.0454	1.0714
0.8	1.0986	1.1270	1.1568	1.1881	1.2212	1.2562	1.2933	1.3331	1.3758	1.4219
0.9	1.4722	1.5275	1.5890	1.6584	1.7380	1.8318	1.9459	2.0923	2.2976	2.6466
0.99	2.6466	2.6996	2.7587	2.8257	2.9031	2.9945	3.1063	3.2504	3.4534	3.8002